FUNDAMENTALS OF MATH

BOOK 1

PRE-ALGEBRA

by

Jerry Ortner

AuthorHouse™
1663 Liberty Drive
Bloomington, IN 47403
www.authorhouse.com
Phone: 1 (800) 839-8640

Published by AuthorHouse 03/03/2015

ISBN: 978-1-4389-9165-8 (sc)
ISBN: 978-1-4969-7177-7 (e)

Print information available on the last page.

This book is printed on acid-free paper.

TABLE OF CONTENTS

LESSON 1 – Place Values

Trillions		Hundred-billions	Ten-billions	Billions		Hundred-millions	Ten-millions	Millions		Hundred-thousands	Ten-thousands	Thousands		Hundreds	Tens	Ones
__	,	__	__	__	,	__	__	__	,	__	__	__	,	__	__	__

Practice on place values.

1. Find the place value of the digit **3** in the number 5,238,876,914,684
    a. billions   b. ten-billions
    c. ten-millions   d. tens

2. Find the place value of the digit **7** in the number 9,238,861,754,184
    a. ten-thousands   b. millions
    c. ten-billions   d. hundred-thousands

3. Find the place value of the digit **2** in the number 8,374,465,129,549
    a. trillions   b. hundreds
    c. millions   d. ten-thousands

4. Find the place value of the digit **7** in the number 5,642,219,873,923
    a. hundred-billions   b. hundred-millions
    c. ten-thousands   d. millions

5. What is the place value of the **7** in 9,235,546,718,658?

6. What is the place value of the **9** in 5,386,694,217,467?

7. What is the place value of the **6** in 8,967,721,435,175?

8. What is the place value of the **7** in 6,354,421,978,148?

9. Find the place value of the digit **3** in the number 6,158,849,327,987
    a. hundred-thousands   b. trillions
    c. millions   d. hundreds

10. Find the place value of the digit **9** in the number 8,654,497,123,743
    a. ten-thousands   b. millions
    c. ten-billions   d. ten-millions

LESSON 2 – Rounding

Look at the question. Decide on what place value the number is to be rounded to and look ***one place to the right*** of that. If less than 5, do not round up and place zeros for each location in the number. If 5 or more, increase the place value by 1 and add zeros for fillers.

Practice on rounding.

1.   Round 9138 to the nearest thousands.
        a. 1000   b. 9100   c. 9000   d. 9140

2.   Round 2663 to the nearest thousands.
        a. 3000   b. 1000   c. 2700   d. 2660

3.   Round 5215 to the nearest thousands.
        a. 1000   b. 5000   c. 5220   d. 5200

4.   Round 45,169 to the nearest tens.

5.   Round 58,276 to the nearest hundreds.

6.   Round 46,482 to the nearest thousands.

7.   Round 1634 to the nearest tens.
        a. 1000   b. 1630   c. 1634   d. 1600

8.   Round 4547 to the nearest hundreds.
        a. 4500   b. 5000   c. 4550   d. 4600

9.   Round 7758 to the nearest thousands.
        a. 8000   b. 7760   c. 7000   d. 7800

10.   Round 57,373 to the nearest thousands.

11.   Round 47,266 to the nearest hundreds.

12.   Round 62,187 to the nearest ten-thousands.

13.   Round 4727 to the nearest tens.

14.   Round 9341 to the nearest hundreds.

15.   Round 1823 to the nearest tens.
                    a. 1830   b. 2000   c. 1800   d. 1820

16.   Round 2752 to the nearest thousands.
                    a. 1000   b. 2800   c. 3000   d. 2750

17.   Round 8141 to the nearest hundreds.
                    a. 8200   b. 8100   c. 8140   d. 8000

18.   Round 68,289 to the nearest thousands.

19.   Round 44,466 to the nearest hundreds.

20.   Round 6,499,990 to the nearest millions.

21.   Round 74,195 to the nearest hundreds.

22.   Round 23,681 to the nearest ten-thousands.
                    a. 23,700   b. 30,000   c. 20,000   d. 24,000

23.   Round 2175 to the nearest tens.
                    a. 2170   b. 2180   c. 2200   d. 1000

24.   Round 4444 to the nearest thousands.
                    a. 4440   b. 5000   c. 4000   d. 4400

25.   Round 64,372 to the nearest hundreds.

26.   Round 86,464 to the nearest thousands.

LESSON 3 – Subtraction of Whole Numbers

There is no problem when the number on the top (the minuend) is equal to or larger than the number on the bottom (the subtrahend).

Examples:

```
    56          147          596         4896
  − 23        − 26        − 345       −2372
    33          121          251         2524
```

However, everything changes when the bottom number is greater than the top.

Example:
```
    52
  − 27

   4  12
   5  2
 − 2  7
   2  5
```

In this problem, you cannot subtract 7 from 2. So you need to borrow a "10" from the 50 and add that 10 to the 2, which now becomes 12 and the 5 reduces to a 4. Now the top number is larger than the bottom, so we may now proceed.

Check your work by adding the answer (the difference) and the minuend.
25 + 27 = 52 ✓ AOK!

Let's try another.

```
    324
  − 158

     1  14
   3  2  4
 − 1  5  8
        6

   2  11  14
   3  2   4
 − 1  5   8
   1  6   6
```

First borrow 10 from the 2 so that the 4 becomes 14, and the 2 becomes a 1.

Now subtract 8 from 14. The result is 6. Moving over one position to the left, we can't subtract 5 from 1. So we have to borrow 10 from the left.

Then 11 − 5 = 6 and finally 2 − 1 = 1.
Check?? Does 166 + 158 = 324?

Try another with zeros on top!

```
    4003
  − 1462
```

First subtract 2 from 3. It can be done.
Result = 1.

4

```
  4   0   0   3
- 1   4   6   2
─────────────────
              1
```

Now we can't subtract 6 from zero (0), so we need to borrow. The next number to the left is another 0. We need to go over one more place to the left. Reduce the 4 to a 3. Place a 10 in the hundreds place. Still 0 – 6. So we need to borrow again.

```
  3   10
  4   0   0   3
- 1   4   6   2
─────────────────
  2   5   4   1
```

The 10 now becomes a 9 and the 0 becomes a 10. Now we have 10 – 4.

```
       9
  3   1̶0̶  10
  4   0   0   3
- 1   4   6   2        9 – 4 = 5   and   3 – 1 = 2
─────────────────
  2   5   4   1
```

Check to make sure the answer is correct:
2541 + 1462 = 4003 ✓

Try another with words rather than numbers.

What number is twelve less than forty?

Note: Less than indicated the subtraction operation.

Rewrite the statement with numbers: 40 – 12 = ⬚ ?

```
  3   10
  4   0          Can't subtract 2 from zero (0), so we need to borrow 10
- 1   2          from 4. It becomes a three and the zero becomes 10.
─────────
  2   8          10 – 2 = 8   and   3 – 1 = 2
```

Check!!   28 + 12 = 40 ✓

Now try these practice problems to see if you can master subtraction.

1.   4000 – 1279   a. 2731   b. 2711   c. 2821   d. 2721

2.   3000 – 1507   a. 1393   b. 1493   c. 1503   d. 1593

3.   What number is twenty-nine less than forty-four?

4.   What number is eighteen less than fifty?

5.   7835 – 243

6.   Subtract: 7000 – 1858

7.   Subtract: 4000 – 1030

8. $6.15
        – $0.98   a. $5.02   b. $6.17   c. $7.13   d. $5.17

9. $5.22
        – $0.98   a. $5.24   b. $6.66   c. $4.24   d. $5.09

10.   Subtract: 8000 – 4333

11.   6526 – 294

12.   5876 – 392

13.   What number is thirty-one less than sixty-six?

14.   4000 – 2757   a. 1143   b. 1253   c. 1343   d. 1243

15.   3000 – 1279   a. 1821   b. 1721   c. 1731   d. 1711

16.   5203 – 2847

LESSON 4 – Multiply and Divide Numbers
Use of calculator is permitted.

1. What is the product of three thousand seven hundred forty-seven and twenty-one?

2. What is the product of two thousand seven hundred twenty-one and seventeen?

3. 126 × 478   a. 59,228   b. 60,128   c. 60,218   d. 60,228

4. 136 × 337   a. 45,822   b. 45,832   c. 44,832   d. 45,732

5. 147 × 209   a. 30,713   b. 30,723   c. 30,623   d. 29,723

6. What is the product of one thousand nine hundred sixty-three and thirty-four?

7. Multiply: 35 × 41

8. Rewrite the addition problem as a multiplication problem and solve. (Different way to express addition.)

    382 + 382 + 382 + 382 + 382 + 382

9.   835
      × 49

10. Multiply: $3.63
              × 60   a. $217.80   b. $435.60   c. $207.80   d. $63.63

11. 23 ⟌ 769   a. 32 R 6   b. 33 R 10   c. 32 R 11   d. 33 R 14

12. 12.   33 ⟌ 536   a. 16 R 22   b. 16 R 8   c. 15 R 18   d. 15 R 31

13. Divide: 3596 ÷ 49

14. Divide: 5476 ÷ 64

15. Divide: $\dfrac{648}{81}$

16. If 132 students are divided into groups of 11, how many groups will be formed?

17. Divide: $\dfrac{957}{33}$

18. Divide: $\dfrac{504}{56}$

19. $3795 \div 87$

20. $1855 \div 72$

21. If 150 players are separated into 15 equal teams, how many players will there be on each team?

22. $6 \overline{) 6291}$

23. Here are three ways to write 72 divided by 8 (Three ways to show division problems.):

$$8 \overline{) 72} \qquad 72 \div 8 \qquad \dfrac{72}{8}$$

Show three ways to write 54 divided by 6.

24. A discount audio store advertised a collection of country music on 11 compact discs for $56.65. What is the cost of each disc?
          a. $5.15   b. $5.26   c. $5.17   d. $5.25

25. If 315 players are separated into 21 equal teams, how many players will there be on each team?

LESSON 5 – Word Problems Focusing on Addition and Subtraction

1. Karl bought a pair of slacks for $33.25 and a shirt for $13.73. What was the total cost of these items?

    a. $46.98   b. $47.98   c. $47.88   d. $47.48

2. Cayla bought a skirt for $41.12 and a blouse for $12.52. What was the total cost of these items?

    a. $52.64   b. $53.64   c. $54.14   d. $52.54

3. What number is twenty-seven less than seventy-nine?

4. Rose paid $10 for a $9.18 book. How much money should she get back?

5. Fred paid $5 for a $1.42 lunch. How much money should he get back?

6. The band club sold 3119 candy bars the first week of the fundraiser to buy new band uniforms. 3365 candy bars were sold the second week. How many candy bars were sold the first two weeks?

7. A soccer stadium holds 24,000 people. If 7000 people can be seated in the bleachers, how many seats are available in the rest of the stadium?

8. The chorus sold 1433 candy bars the first week of the fundraiser. 5994 candy bars were sold the second week. How many candy bars were sold the first two weeks?

    a. 7427 candy bars   b. 7027 candy bars
    c. 7407 candy bars   d. 7424 candy bars

9. A football stadium holds 17,000 people. If 5000 people can be seated in the bleachers, how many seats are available in the rest of the stadium?

10. Carl bought a pair of sneakers for $64.00 and socks for $12.50. What was the total cost of these items?

    a. $51.50   b. $64.00   c. $76.50   d. $12.50

LESSON 6 – Add, Subtract, and Multiply Decimals

When adding decimals, keep the decimal points in a straight line (vertically). You can fill in zeros if you wish. Then add each column, starting at the right and moving left. Likewise, when subtracting, decimal beneath decimal.

Example 1: 2.3 + 4.363 + 5.12 + 7 + 8.132

```
      2.300
      4.630
      5.120
      7.000
   +  8.132
     27.182
```

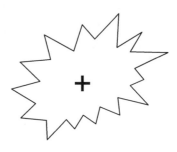

Example 2: 24.02 – 15.147

```
        13    9  11
    1    3   10  4   10
    2    4 . 0   2   0
 -  1    5 . 1   4   7
         8 . 8   7   3
```

OH!! Don't forget the decimal point in your answer!

When multiplying, count off the places in the answer.

Example 3: 16.24 × 1.3

```
      1  6 . 2  4   ←— 2 decimal places ⎫ total of 3 decimal
   ×        1 . 3   ←— 1 decimal place  ⎬ places
      4  8  7  2
   1  6  2  4
   2  1. 1  1  2.   ←— starting from the far right, move the
                        decimal point 3 places to the left.
```

Therefore, 16.24 × 1.3 = 21.112.

Practice problems on adding, subtracting, and multiplying decimals.

1. $2.8 + 0.3 + 2.3$ a. 12.4   b. 10.4   c. 4.9   d. 5.4

2. $0.6 + 0.32 + 5.32$ a. 5.94   b. 6.24   c. 10.24   d. 15.24

3. Add: $1.84 + 2.22 + 4.71$

4. Subtract: $39.4 - 7.32$

5. $34.3 - 5.95$

6. $9.81 + 8 + 8.1$ a. 25.92   b. none of these   c. 26.91   d. 26.01

7. $30.7 - 18$ a. 28.9   b. 1.27   c. 12.7   d. 289.0

8. $2.48 + 8 + 4.9$

9. What is the sum of 2.18, 0.465, and 6? (sum means to add)

10. 0.76×0.01

11. 0.56×0.06

12. 0.89×0.07 a. 0.00623   b. 6.23   c. 0.623   d. 0.0623

13. $0.5 \times 0.6 \times 0.3$

14. $0.6 \times 0.2 \times 0.7$

15. 0.86×0.06

16. 0.88×0.09

17. $7 - (2.29 + 0.61)$ add the two numbers inside the parentheses before subtracting

LESSON 7 – Dividing Decimals Using Whole Numbers as Divisors

Some examples to start.

Example 1:   3.84 ÷ 4

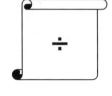

If the divisor is a whole number, place the decimal directly above its location in the dividend.

```
        .  9  6
      ↑
  4 | 3  .  8  4
      3     6
            2  4
            2  4
               0
```

Example 2:   .0245 ÷ 7

Make certain the placeholders (zeros) are inserted to the right of the decimal point when you can't divide.

```
     .  0  0  3  5
  7 | .  0  2  4  5
           2  1
              3  5
              3  5
                 0
```

Example 3:   256.44 ÷ 24

Add zeros to the dividend so the problem can be completed.

```
        1  0  .  6  8  5
  24 | 2  5  6  .  4  4  0
       2  4
       1  6     4
       1  4     4
          2     0  4
          1     9  2
                1  2  0
                1  2  0
                      0
```

12

Practice problems on dividing decimals.

1. $0.84 \div 4$

2. $9 \overline{)6.3}$

3. $6 \overline{)0.36}$

4. $2.46 \div 6$

5. $2.13 \div 3$

6. $4 \overline{)0.24}$ a. 0.06   b. 0.9   c. 0.09   d. 0.6

7. $5 \overline{)0.35}$ a. 0.7   b. 0.9   c. 0.09   d. 0.07

8. $1.68 \div 4$

9. $4.97 \div 7$

10. $1.26 \div 6$

11. $2.05 \div 5$

12. $1.28 \div 4$

13. $3 \overline{)0.12}$ a. 0.4   b. 0.04   c. 0.06   d. 0.6

14. $7 \overline{)2.1}$ a. 0.03   b. 0.02   c. 0.2   d. 0.3

15. $7\overline{)0.35}$ a. 0.5   b. 0.7   c. 0.05   d. 0.07

16. $5\overline{)0.3}$ a. 0.06   b. 0.7   c. 0.6   d. 0.07

17. $4 \div 2000$

18. $6\overline{)0.36}$

19. $9.2 \div 4$

20. $8\overline{)0.736}$

21. $32\overline{)12}$

22. $91.44 \div 36$

23. $2\overline{)5.328}$

24. $121.44 \div 24$

25. $48\overline{)60}$

LESSON 8 – Converting Decimals to Fractions and Vice Versa

Here again place value comes in handy. Go back to Lesson 1 to review place values.

To the right of the decimal point, the place values are as follows:

. | tenths | hundredths | thousandths | ten-thousandths

Several examples converting a decimal to a common fraction:

Example 1: $0.3 = \dfrac{3}{10}$

Example 2: $0.46 = \dfrac{46}{100}$ (which reduces to $\dfrac{23}{50}$)

Example 3: $0.571 = \dfrac{571}{1000}$

A good way to check is to look at the denominator of the fraction. It should have as many zeros as places to the right of the decimal point. In example 1, there is one place to the right of the decimal point so there will be one zero in the denominator of the fraction. Example 2, two places so two zeros, etc.

Also remember to be able to work with a whole number and decimal.

Example 4: $23.67 = 23\dfrac{67}{100}$

Example 5: $1234.56789 = 1234\dfrac{56789}{100,000}$

Converting a fraction to a decimal is the reverse operation. Count the number of zeros in the denominator. That is, how many zeros there are after the number 1. (Note: make certain that the denominator is in tenths, hundredths, thousandths, etc)

Example 6:   What is $\frac{72}{100}$ as a decimal? Answer: 0.72

Example 7:   What is $1\frac{37}{1000}$ as a decimal? Answer: 1.037

Example 8:   What is $\frac{9}{10}$ as a decimal? Answer: 0.9

Example 9:   What is $\frac{125}{500}$ as a decimal? Answer: 0.25

Convert: $\frac{125}{500} = \frac{x}{1000}$ cross-multiply to solve: x = 250
(Note: the zero after the 5 is not necessary in the decimal answer.)

Now try the practice problems and see if you understand this lesson thoroughly.

1.   Write 0.59 as a common fraction.
          a. $\frac{590}{100}$ b. $\frac{5.9}{100}$ c. $\frac{59}{100}$ d. $\frac{59}{10}$

2.   Write 0.45 as a common fraction.
          a. $\frac{4.5}{100}$ b. $\frac{45}{100}$ c. $\frac{45}{10}$ d. $\frac{450}{100}$

3.   Write 0.83 as a common fraction.

4.   Write 0.27 as a common fraction.

5.   Write $\frac{50}{100}$ as a decimal number.

6.   Write 0.11 as a common fraction.

7.   Write 0.47 as a common fraction.

8.   Write 0.65 as a common fraction.
          a. $\frac{6.5}{100}$ b. $\frac{650}{100}$ c. $\frac{65}{10}$ d. $\frac{65}{100}$

9.   What is $\frac{33}{1000}$ as a decimal?
          a. 0.0033   b. 0.033   c. 3.3   d. 0.33

10.   What is $\frac{21}{100}$ as a decimal?
          a. 2.1   b. 0.0021   c. 0.21   d. 0.021

11. What is $\frac{27}{1000}$ as a decimal?

            a. 2.7   b. 0.0027   c. 0.27   d. 0.027

12. What is $\frac{3}{10}$ as a decimal?

            a. 0.003   b. 0.03   c. 0.3   d. 0.0003

13. Write 0.39 as a common fraction.

            a. $\frac{3.9}{100}$ b. $\frac{390}{100}$ c. $\frac{39}{10}$ d. $\frac{39}{100}$

14. Write 0.33 as a common fraction.

            a. $\frac{3.3}{100}$ b. $\frac{33}{100}$ c. $\frac{330}{100}$ d. $\frac{33}{10}$

15. Write 0.79 as a common fraction.

16. Write 0.29 as a common fraction.

17. Write 0.75 as a common fraction.

            a. $\frac{75}{100}$ b. $\frac{7.5}{100}$ c. $\frac{75}{10}$ d. $\frac{750}{100}$

18. Write 0.25 as a common fraction.

            a. $\frac{250}{100}$ b. $\frac{2.5}{100}$ c. $\frac{25}{10}$ d. $\frac{25}{100}$

19. Write 0.43 as a common fraction.

            a. $\frac{4.3}{100}$ b. $\frac{43}{10}$ c. $\frac{43}{100}$ d. $\frac{430}{100}$

20. What is $\frac{16}{1000}$ as a decimal?

            a. 0.16   b. 0.016   c. 0.0016   d. 1.6

21. What is $\frac{4}{10}$ as a decimal?

            a. 0.004   b. 0.0004   c. 0.04   d. 0.4

22. Write 0.77 as a common fraction.

            a. $\frac{77}{100}$ b. $\frac{77}{10}$ c. $\frac{7.7}{100}$ d. $\frac{770}{100}$

23. Write 0.55 as a common fraction.

            a. $\frac{55}{10}$ b. $\frac{5.5}{100}$ c. $\frac{55}{100}$ d. $\frac{550}{100}$

24. Write 0.73 as a common fraction.

25. Write 0.37 as a common fraction.

26. Write $\frac{70}{100}$ as a decimal number.

LESSON 9 – Divisibility Tests

The Rules for Divisibility are:

divisible by 2	even number
divisible by 3	the sum of the digits is divisible by 3
divisible by 4	last 2 digits are a multiple of 4, divide last two digits by 4, if no remainder, then divisible by 4
divisible by 5	last digit is a 0 or 5
divisible by 6	the sum of the digits is divisible by 3 AND an even number
divisible by 8	the last three digits of the number is divisible by 8
divisible by 9	the sum of the digits is divisible by 9
divisible by 10	last digit is a 0

In certain textbooks, divisibility is shown as 3|1568. Is 1568 divisible by 3? Another example: 6|8142. Is 8142 divisible by 6? The answer to these are: 3|1568 is no and 6|8142 is yes.

Some practice problems to try.

1.   Is 426 divisible by both 5 and 10?

2.   Which of these numbers is divisible by both 9 and 2?
          a. 126   b. 567   c. 117   d. 22

3.   Which of these numbers is divisible by both 10 and 3?
          a. 1301   b. 71   c. 390   d. 52

4.   Which of these numbers is divisible by both 3 and 9?
          a. 40   b. 62   c. 189   d. 98

5.   Which of these numbers is divisible by both 9 and 10?
          a. 64   b. 1170   c. 1054   d. 171

6.   Which of these numbers is divisible by both 10 and 5?
          a. 1701   b. 111   c. 176   d. 850

7.   Which of these numbers is divisible by both 3 and 2?
          a. 66   b. 51   c. 99   d. 52

8.   Is 116 divisible by both 3 and 9?

9.   Is 390 divisible by both 10 and 3?

10.   Is 566 divisible by both 9 and 10?

11.   Is 32 divisible by both 3 and 10?

12.   Is 195 divisible by both 5 and 3?

13.   Which of these numbers is divisible by both 3 and 5?
                                      a. 153   b. 33   c. 175   d. 255

14.   Which of these numbers is divisible by both 9 and 3?
                                        a. 566   b. 189   c. 34   d. 118

15.   Which of these number is divisible by both 10 and 2?
                                        a. 220   b. 27   c. 171   d. 1101

16.   Which of these numbers is divisible by both 2 and 10?
                                        a. 15   b. 53   c. 260   d. 171

17.   Is 117 divisible by both 3 and 2?

18.   Is 129 divisible by both 5 and 10?

19.   Is 109 divisible by both 10 and 9?

20.   Is 70 divisible by both 2 and 5?

21.   4|15,984

22.   5|38,814

23.   6|104,538

24.   8|28,096

25.   9|11,378

26.   3|5958

27.   4|10,612

28.   5|48,670

29.   9|23,772

30.   6|163,944

LESSON 10 – Prime and Composite Numbers

By definition, a prime number has only itself and one as its factors. All others are composite. The prime numbers less than 20 are 2, 3, 5, 7, 11, 13, 17, and 19.

Practice problems on prime and composite numbers.

1.   Which of the numbers is prime: 21, 57, 39, 31

2.   Which of the numbers is prime: 15, 35, 83, 25

3.   Which of the following is a prime number?
                a. 27   b. 25   c. 67   d. 33

4.   Which of the following is a prime number?
                a. 21   b. 73   c. 39   d. 15

5.   Which of the numbers is prime: 57, 73, 25, 27

6.   Which of the numbers is prime: 35, 51, 47, 21

7.   Which of the following is a prime number?
                a. 27   b. 89   c. 9   d. 57

8.   Which of the following is a prime number?
                a. 29   b. 33   c. 35   d. 21

9.   Write the composite numbers from the list: 19, 29, 18, 37, 21, 26, 43, 47, 10

10.   Write the composite numbers from the list: 5, 12, 13, 16, 19, 18, 29, 6, 37

11.   How many of the following numbers are composite: 12, 36, 14, 4, 32, 55
                a. 5   b. none   c. 4   d. 6

12. How many of the following numbers are composite: 28, 43, 24, 22, 38, 18, 8, 54
    a. 5   b. none   c. 8   d. 7

13. How many of the following numbers are composite: 42, 51, 3, 26, 15, 52, 19, 33
    a. 8   b. 6   c. 5   d. 3

14. Write the composite numbers from the list: 29, 37, 43, 22, 26, 47, 2, 32, 12

15. How many of the following numbers are composite: 31, 22, 20, 24, 15, 41, 3, 28
    a. 5   b. none   c. 8   d. 4

16. List the first 10 composite numbers.

17. List all composite numbers between 20 and 30.

18. List all prime numbers between 20 and 50.

19. Which of these numbers are composite: 8, 35, 47, 77, 67, 54, 33, 71, 19, 57

20. What are the composite numbers between 60 and 70?

21. How many prime numbers are there between 12 and 25? What are they?

22. True or False: When you add two prime numbers, the sum is always an even number.

23. How many single-digit composite numbers are there?

24. List the single-digit numbers.

LESSON 11 – Prime Factorization of a Number

Having found the first ten prime numbers, we can move forward with factorization of any number.

Let's take this example: 216. This is how I figure out prime factors of a number:

$$
2^3 \begin{cases} 2 & | & 216 \\ 2 & | & 108 \\ 2 & | & 54 \end{cases}
$$
$$
3^3 \begin{cases} 3 & | & 27 \\ 3 & | & 9 \\ 3 & | & 3 \\ & & 1 \end{cases}
$$

We finally get $2^3 \cdot 3^3$. Both 2 and 3 are prime numbers. Therefore the prime factorization of 216 is $2^3 \cdot 3^3$.

Try another with me before you tackle the problems below. The number is 180.

$$
2^2 \begin{cases} 2 & | & 180 \\ 2 & | & 90 \end{cases}
$$
$$
3^2 \begin{cases} 3 & | & 45 \\ 3 & | & 15 \end{cases}
$$
$$
5 \begin{cases} 5 & | & 5 \\ & & 1 \end{cases}
$$

The prime factorization of 180 is $2^2 \cdot 3^2 \cdot 5$.

Try these practice problems. If the answer is given, multiply the prime numbers to ascertain its product. Good luck!

1. Determine the prime factorization of 2100.
    a. $2 \cdot 2 \cdot 3 \cdot 5 \cdot 5 \cdot 7$ b. $2 \cdot 2 \cdot 2 \cdot 3 \cdot 5 \cdot 7$
    c. $2 \cdot 2 \cdot 3 \cdot 3 \cdot 5 \cdot 5 \cdot 11$ d. $2 \cdot 3 \cdot 5 \cdot 10 \cdot 11$

2. Determine the prime factorization of 990.
    a. $2 \cdot 3 \cdot 3 \cdot 3 \cdot 5 \cdot 7$ b. $3 \cdot 3 \cdot 5 \cdot 5 \cdot 7 \cdot 10$
    c. $2 \cdot 2 \cdot 3 \cdot 3 \cdot 5 \cdot 5 \cdot 11$ d. $2 \cdot 3 \cdot 3 \cdot 5 \cdot 11$

3. Write 700 as a product of prime numbers.

4. Write 525 as a product of prime numbers.

5. Write 1260 as a product of prime numbers.

6. Determine the prime factorization of 2200.

   a. 2 • 2 • 2 • 3 • 5 • 5 • 7   b. 2 • 2 • 2 • 2 • 5 • 11
   c. 2 • 2 • 2 • 5 • 5 • 11   d. 2 • 2 • 5 • 7 • 10

7. Determine the prime factorization of 630.

   a. 2 • 3 • 3 • 3 • 5 • 11   b. 2 • 2 • 3 • 3 • 5 • 7
   c. 2 • 2 • 2 • 5 • 5 • 11   d. 2 • 2 • 5 • 7 • 10

8. Determine the prime factorization of 440.

   a. 2 • 2 • 2 • 5 • 11   b. 2 • 2 • 2 • 2 • 5 • 5 • 11
   c. 2 • 2 • 5 • 5 • 7 • 10   d. 2 • 2 • 2 • 3 • 5 • 7

9. Write 4200 as a product of prime numbers.

10. Write 9900 as a product of prime numbers.

11. Write 420 as a product of prime numbers.

12. Determine the prime factorization of 3150.

   a. 2 • 3 • 3 • 3 • 5 • 5 • 11   b. 3 • 3 • 5 • 10 • 11
   c. 2 • 2 • 3 • 3 • 5 • 7   d. 2 • 3 • 3 • 5 • 5 • 7

13. Determine the prime factorization of 1320.

   a. 2 • 2 • 2 • 3 • 3 • 5 • 7   b. 2 • 2 • 2 • 3 • 5 • 11
   c. 2 • 2 • 3 • 5 • 5 • 7 • 10   d. 2 • 2 • 2 • 2 • 3 • 5• 5•11

14. Determine the prime factorization of 1100.

   a. 2 • 5 • 7 • 10   b. 2 • 2 • 5 • 5 • 11
   c. 2 • 2 • 3 • 5 • 5 • 7   d. 2 • 2 • 2 • 5 • 11

15. Write 440 as a product of prime numbers.

16. Write 19,800 as a product of prime numbers.

17. Write 6300 as a product of prime numbers.

18. Determine the prime factorization of 6600.

   a. 2 • 2 • 2 • 2 • 3 • 5 • 11   b. 2 • 2 • 2 • 3 • 5 • 5 • 11
   c. 2 • 2 • 2 • 3 • 3 • 5 • 5 • 7   d. 2 • 2 • 3 • 5 • 7 • 10

19. Determine the prime factorization of 1400.

   a. 2 • 2 • 2 • 3 • 5 • 5 • 11   b. 2 • 2 • 5 • 10 • 11
   c. 2 • 2 • 2 • 2 • 5 • 7   d. 2 • 2 • 2 • 5 • 5 • 7

20. Determine the prime factorization of 330.

   a. 2 • 2 • 3 • 5 • 5 • 11   b. 2 • 3 • 5 • 11
   c. 2 • 3 • 3 • 5 • 7   d. 3 • 5 • 5 • 7 • 10

LESSON 12 – Multiples

Example: List the first ten multiples of 2. It is the 2's times table.

2, 4, 6, 8, 10, 12, 14, 16, 18, 20

Practice problems on multiples.

1.   List the first five multiples of 6.
                  a. 6, 12, 18, 24, 30   b. 30, 35, 40, 45, 50
                  c. 6, 7, 8, 9, 10   d. 0, 1, 6, 12, 18

2.   List the first five multiples of 5.
                  a. 0, 1, 5, 10, 15   b. 5, 6, 7, 8, 9
                  c. 5, 10, 15, 20, 25   d. 25, 30, 35, 40, 45

3.   List the first five multiples of 4.
                  a. 4, 8, 12, 16, 20   b. 20, 25, 30, 35, 40
                  c. 0, 1, 4, 8, 12   d. 4, 5, 6, 7, 8

4.   List the first five multiples of 3.
                  a. 3, 6, 9, 12, 15   b. 0, 1, 3, 6, 9
                  c. 15, 20, 25, 30, 35   d. 3, 4, 5, 6, 7

5.   List the first five multiples of 7.
                  a. 7, 14, 21, 28, 35   b. 0, 1, 7, 14, 21
                  c. 7, 8, 9, 10, 11   d. 35, 40, 45, 50, 55

6.   List the first five multiples of 9.
                  a. 45, 50, 55, 60, 65   b. 9, 18, 27, 36, 45
                  c. 0, 1, 9, 18, 27   d. 9, 10, 11, 12, 13

7.   List the first five multiples of 8.
                  a. 8, 9, 10, 11, 12   b. 40, 45, 50, 55, 60
                  c. 8, 16, 24, 32, 40   d. 0, 1, 8, 16, 24

8.   List the first six multiples of 10.

9.   List the first five multiples of 12.

10.   List the first nine multiples of 11.

LESSON 13 – Least Common Multiple

There are two distinct ways to find the least common multiple (LCM). The LCM means the smallest number that is divisible by **_all_** the numbers.

The first way to find the LCM is to list the sets of multiples of all numbers. Use 12 and 16 as an example.

Multiples of 12: {12, 24, 36, 48, 60, 72, 84, 96, …}
Multiples of 16: {16, 32, 48, 64, 80, 96, …}

Upon inspection, the first two common multiples of 12 and 16 are 48 and 96. The smallest of these is 48.

The other way to find the LCM is (1) prime factor each number, (2) select every prime factor raised to the highest power and (3) find their product. Again, using 12 and 16.

Step 1: 12 prime factors into 2^2 x 3
16 prime factors into 2^4

Step 2: 2^4 and 3

Step 3: 2^4 x 3 = 48, the least common multiple

Now try these for good luck.

1. Find the least common multiple of 8 and 4.
a. 32   b. 8   c. 12   d. 4

2. Find the least common multiple of 6 and 2.
a. 6   b. 12   c. 8   d. 2

3. What is the least common multiple of 2 and 8?

4. Find the least common multiple of 6, 2, and 12.

5. Find the least common multiple of 2, 10, and 3.

6. Find the least common multiple of 3, 6, and 8.

7. Find the least common multiple of 8, 2, and 9.

8. What is the least common multiple of 2 and 10?

9. What is the least common multiple of 9 and 6?

10. Find the least common multiple of 10 and 8.
    a. 80   b. 2   c. 40   d. 18

11. What is the least common multiple of 6 and 8?

12. Find the least common multiple of 2, 4, and 6.

13. What is the least common multiple of 5 and 7?

14. Find the least common multiple of 3, 4, and 5.

15. Find the least common multiple of 2, 4, and 3.

16. What is the least common multiple of 10 and 5?

17. Find the least common multiple of 8 and 12.
    a. 4   b. 96   c. 24   d. 20

18. Find the least common multiple of 10, 6, and 4.

19. Find the least common multiple of 2, 3, and 5.

20. Find the least common multiple of 3, 5, and 9.

21. What is the least common multiple of 4, 6, and 8?

22. Find the least common multiple of 8, 10, and 12.

23. Find the least common multiple of 24, 12, and 6.

24. What is the least common multiple of 5, 10, and 25?

25. Find the least common multiple of 6, 8, and 9.

LESSON 14 – Exponents

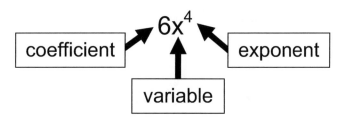

Example 1: $x^3 = x \cdot x \cdot x$

Example 2: $4^3 = 4 \cdot 4 \cdot 4 = 64$

Example 3: $a^2 = a \cdot a$

Example 4: $3b^3 = 3 \cdot b \cdot b \cdot b$

Example 5: $2^5 = 2 \cdot 2 \cdot 2 \cdot 2 \cdot 2 = 32$

Example 6: $(-3)^3 = (-3) \cdot (-3) \cdot (-3) = (+9) \cdot (-3) = -27$

Practice on exponents. Evaluate the following.

1. 2^4 a. 17   b. 16   c. 15   d. 18

2. 3^4 a. 12   b. 81   c. 243   d. 64

3. 2^3

4. 5^2

5. 4^2

6. 4^3

7. 5^3

8. 7^2

9. 3^3 a. 26   b. 29   c. 28   d. 27

10. 3^2 a. 9   b. 6   c. 27   d. 8

11. 2^3 a. 16   b. 9   c. 8   d. 6

12. 6^2

13. 8^2

14. 7^3

15. 10^2

16. 9^2

17. 4^4

18. 2^6

19. 8^3

20. 9^3

21. 10^3

22. 5^4

23. 11^2

24. 15^2

25. 12^2

26. 13^2

27. 2^5

28. $(-3)^5$

29. 6^3

30. $(-2x)^2$

31. $(-5x)^3$

32. $(10a)^2$

33. $2(ay)^3$

LESSON 15 – Greatest Common Factor

The **G**reatest **C**ommon **F**actor (**GCF**) of two numbers is the greatest whole number that is a factor of each number. To find the GCF, find the prime factorization of each number and then find the greatest power of each factor common to BOTH numbers. Their product is the GCF.

Example 1:   Find the GCF of $\frac{16}{72}$.

$$16 = 2 \cdot 2 \cdot 2 \cdot 2 = \boxed{2^3} \cdot 2$$
$$72 = 2 \cdot 2 \cdot 2 \cdot 3 \cdot 3 = \boxed{2^3} \cdot 3^3$$

2^3 is common to both numbers.
Therefore the GCF is $2^3 = 8$.

Example 2:   Find the GCF of $\frac{49}{63}$.

$$49 = \boxed{7} \cdot 7$$
$$63 = 7 \cdot 3 \cdot 3 = \boxed{7} \cdot 3^2$$

7 is common to both numbers.
Therefore the GCF is 7.

Example 3:   Find the GCF of $\frac{144}{256}$.

$$144 = 2 \cdot 2 \cdot 2 \cdot 2 \cdot 3 \cdot 3 = \boxed{2^4} \cdot 3^2$$
$$256 = 2 \cdot 2 \cdot 2 \cdot 2 \cdot 2 \cdot 2 \cdot 2 \cdot 2 = \boxed{2^4} \cdot 2^4$$

2^4 is common to both numbers. Therefore the GCF is $2^4 = 16$.

Example 4:   Find the GCF of 36 and 78.

$$36 = 2 \cdot 2 \cdot 3 \cdot 3 = \boxed{2 \cdot 3} \cdot 2 \cdot 3$$
$$78 = \boxed{2 \cdot 3} \cdot 13$$

$2 \cdot 3$ is common to both numbers. Therefore the GCF is $2 \cdot 3 = 6$.

Example 5:   Find the GCF of 48 and 120.

$$48 = 2 \cdot 2 \cdot 2 \cdot 2 \cdot 3 = \boxed{2^3 \cdot 3} \cdot 2$$
$$120 = 2 \cdot 2 \cdot 2 \cdot 3 \cdot 5 = \boxed{2^3 \cdot 3} \cdot 5$$

$2^3 \cdot 3$ is common to both numbers. Therefore the GCF is $2^3 \cdot 3 = 24$.

Practice problems on finding the GCF. Find the GCF for the following.

1. 24 and 32

2. 10 and 15

3. 18 and 45

4. 16 and 64

5. 24 and 42

6. 32 and 80

7. 60 and 108

8. 112 and 128

9. 342 and 380

10. 210 and 252

11. 308 and 418

12. 144 and 300

13. 30 and 48

14. 36 and 150

15. 216 and 254

16. 48 and 72

17. 72 and 120

18. 60 and 108

19. 66 and 90

20. 150 and 480

LESSON 16 – Reducing Fractions

Now we can use the LCM and GCF in working with fractions. To reduce a fraction, find the GCF for both numerator and denominator simultaneously. Divide the GCF into both top and bottom of the fraction. The result is a fraction in lowest (reduced) form. Some examples are:

Example 1: $\dfrac{36}{48} = \dfrac{2^2 \cdot 3^2}{2^4 \cdot 3} = \dfrac{\mathbf{12} \cdot 3}{\mathbf{12} \cdot 4} = \dfrac{3}{4}$

Example 2: $\dfrac{40}{50} = \dfrac{2^3 \cdot 5}{2 \cdot 5^2} = \dfrac{\mathbf{10} \cdot 4}{\mathbf{10} \cdot 5} = \dfrac{4}{5}$

Example 3: $\dfrac{18}{64} = \dfrac{2 \cdot 3^2}{2 \cdot 2^5} = \dfrac{\mathbf{2} \cdot 9}{\mathbf{2} \cdot 32} = \dfrac{9}{32}$

Example 4: $\dfrac{27}{42} = \dfrac{3^3}{2 \cdot 3 \cdot 7} = \dfrac{\mathbf{3} \cdot 9}{\mathbf{3} \cdot 14} = \dfrac{9}{14}$

Example 5: $\dfrac{12}{60} = \dfrac{2^2 \cdot 3}{2^2 \cdot 3 \cdot 5} = \dfrac{\mathbf{12} \cdot 1}{\mathbf{12} \cdot 5} = \dfrac{1}{5}$

Example 6: $\dfrac{15}{36} = \dfrac{3 \cdot 5}{2^2 \cdot 3^2} = \dfrac{\mathbf{3} \cdot 5}{\mathbf{3} \cdot 12} = \dfrac{5}{12}$

Example 7: $\dfrac{56}{91} = \dfrac{2^3 \cdot 7}{7 \cdot 13} = \dfrac{\mathbf{7} \cdot 8}{\mathbf{7} \cdot 13} = \dfrac{8}{13}$

Practice on reducing fractions. Reduce to lowest terms.

1. $\dfrac{27}{45}$ a. $\dfrac{3}{5}$ b. $\dfrac{2}{9}$ c. $\dfrac{3}{4}$ d. $\dfrac{9}{15}$

2. $\dfrac{60}{80}$ a. $\dfrac{15}{20}$ b. $\dfrac{4}{5}$ c. $\dfrac{3}{4}$ d. $\dfrac{6}{8}$

3. $\dfrac{12}{30}$ a. $\dfrac{1}{2}$ b. $\dfrac{4}{10}$ c. $\dfrac{2}{5}$ d. $\dfrac{4}{7}$

4. $\dfrac{36}{48}$ a. $\dfrac{5}{36}$ b. $\dfrac{3}{4}$ c. $\dfrac{6}{7}$ d. $\dfrac{9}{12}$

5. $\dfrac{48}{60}$ a. $\dfrac{8}{9}$ b. $\dfrac{1}{8}$ c. $\dfrac{4}{5}$ d. $\dfrac{16}{20}$

6. $\dfrac{16}{24}$ a. $\dfrac{4}{5}$ b. $\dfrac{2}{3}$ c. $\dfrac{4}{6}$ d. $\dfrac{1}{4}$

7. $\frac{18}{36}$ a. $\frac{6}{12}$ b. $\frac{1}{2}$ c. $\frac{5}{7}$ d. $\frac{5}{18}$

8. $\frac{64}{80}$ a. $\frac{4}{5}$ b. $\frac{16}{20}$ c. $\frac{3}{32}$ d. $\frac{8}{9}$

9. $\frac{18}{72}$ a. $\frac{1}{5}$ b. $\frac{1}{6}$ c. $\frac{1}{4}$ d. $\frac{5}{18}$

10. $\frac{108}{144}$ a. $\frac{5}{36}$ b. $\frac{3}{4}$ c. $\frac{9}{12}$ d. $\frac{6}{7}$

11. $\frac{48}{54}$ a. $\frac{4}{5}$ b. $\frac{1}{8}$ c. $\frac{8}{9}$ d. $\frac{16}{20}$

12. $\frac{24}{40}$

13. $\frac{60}{72}$

14. $\frac{28}{70}$

15. $\frac{24}{32}$

16. $\frac{18}{27}$ a. $\frac{2}{9}$ b. $\frac{2}{3}$ c. $\frac{5}{6}$ d. $\frac{6}{9}$

17. $\frac{24}{40}$ a. $\frac{5}{7}$ b. $\frac{6}{10}$ c. $\frac{3}{5}$ d. $\frac{1}{4}$

18. $\frac{32}{80}$ a. $\frac{3}{16}$ b. $\frac{8}{20}$ c. $\frac{2}{3}$ d. $\frac{2}{5}$

19. $\frac{18}{30}$ a. $\frac{5}{7}$ b. $\frac{3}{5}$ c. $\frac{1}{3}$ d. $\frac{6}{10}$

20. $\frac{65}{100}$

21. $\frac{25}{75}$

22. $\frac{16}{40}$

23. $\frac{66}{99}$

24. $\frac{28}{32}$

25. $\frac{15}{50}$

26. $\frac{15}{25}$

27. $\frac{27}{81}$

28. $\frac{20}{48}$

29. $\frac{12}{72}$

30. $\frac{35}{75}$

31. $\frac{42}{48}$

32. $\frac{34}{51}$

33. $\frac{9}{36}$

34. $\frac{12}{30}$

35. $\frac{27}{54}$

36. $\frac{15}{40}$

37. $\frac{24}{60}$

38. $\frac{36}{54}$

39. $\frac{70}{100}$

40. $\frac{90}{360}$

LESSON 17 – Reciprocals

Two numbers, $\dfrac{a}{b}$ and $\dfrac{b}{a}$, whose product is 1 are reciprocals

$$\tfrac{6}{11} \times \boxed{n} = 1 \;\rightarrow\; n = \tfrac{11}{6}$$

$$7 \times n = 1 \;\rightarrow\; n = \tfrac{1}{7}$$

Another name for the reciprocal is **multiplicative inverse**.

The multiplicative inverse of 5 is $\tfrac{1}{5}$.

The multiplicative inverse of $-\tfrac{1}{6}$ is -6.

Practice on reciprocals.

1. a. What is the reciprocal of -9?
    b. What is the reciprocal of $\tfrac{1}{9}$?

2. Which real number does not have a reciprocal and why?

3. What is the multiplicative inverse of $-\tfrac{1}{7}$?

    a. 7   b. -7 c. $-\tfrac{1}{7}$ d. $\tfrac{1}{7}$

4. What is the reciprocal of $\tfrac{1}{8}$?

    a. $-\tfrac{1}{8}$ b. $\tfrac{1}{8}$ c. 8   d. -8

5. What is the multiplicative inverse of -3?

    a. -3 b. $-\tfrac{1}{3}$ c. $\tfrac{1}{3}$ d. 3

6. a. What is the multiplicative inverse of $\tfrac{1}{8}$?
    b. What is the multiplicative inverse of -8?

7.   a.   What is the reciprocal of $-\frac{1}{2}$?
      b.   What is the reciprocal of 2?

8.   What is the reciprocal of 8?

        a. 8   b. -8 c. $-\frac{1}{8}$ d. $\frac{1}{8}$

9.   What is the multiplicative inverse of -7?

        a. $\frac{1}{7}$ b. -7 c. $-\frac{1}{7}$ d. 7

10.   What is the reciprocal of $-\frac{1}{5}$?

        a. -5 b. $-\frac{1}{5}$ c. 5   d. $\frac{1}{5}$

11.   What is the multiplicative inverse of 4?

        a. -4 b. $\frac{1}{4}$ c. $-\frac{1}{4}$ d. 4

12.   What is the multiplicative inverse of -2?

        a. -2 b. $-\frac{1}{2}$ c. 2   d. $\frac{1}{2}$

LESSON 18 – Multiplying and Dividing Fractions

The rule for multiplication of fractions is $\dfrac{a}{b} \cdot \dfrac{c}{d} = \dfrac{ac}{bd}$, where "b" and "d" are not equal to zero.

Example 1: $\quad \dfrac{2}{3} \times \dfrac{4}{5} = \dfrac{2 \times 4}{3 \times 5} = \dfrac{8}{15}$

Example 2: $\quad \dfrac{7}{8} \cdot \dfrac{4}{7} = \dfrac{{}^{1}\!7 \cdot 4}{8 \cdot {}_{1}7} = \dfrac{1 \cdot 4}{8 \cdot 1} = \dfrac{4}{8} = \dfrac{1}{2}$

1. Remember to cancel when the same number is in the top and bottom.
2. Also, reduce fraction to lowest terms.

Example 3: $\quad \dfrac{5}{12} \cdot \dfrac{4}{10} = \dfrac{{}^{1}\!5 \cdot {}^{1}\!4}{{}_{3}12 \cdot {}_{2}10} = \dfrac{1}{3 \cdot 2} = \dfrac{1}{6}$

Example 4: $\quad \dfrac{4}{9} \times \dfrac{10}{15} = \dfrac{4 \times {}^{2}10}{9 \times {}_{3}15} = \dfrac{4 \times 2}{9 \times 3} = \dfrac{8}{27}$

The rule for division of fractions is to multiply by the reciprocal of the **divisor**.

$$\dfrac{a}{b} \div \dfrac{\mathbf{c}}{\mathbf{d}} = \dfrac{a}{b} \cdot \dfrac{d}{c} = \dfrac{ad}{bc} \qquad \text{where "b", "c", and "d"} \neq 0$$

$$\underset{\textbf{divisor}}{\uparrow} \qquad \underset{\textbf{reciprocal}}{\uparrow}$$

Example 5: $\quad \dfrac{4}{7} \div \dfrac{9}{20} = \dfrac{4}{7} \cdot \dfrac{20}{9} = \dfrac{4 \cdot 20}{7 \cdot 9} = \dfrac{80}{63} = 1\dfrac{17}{63}$

Example 6: $\quad \dfrac{3}{2} \div \dfrac{3}{8} = \dfrac{3}{2} \cdot \dfrac{8}{3} = \dfrac{{}^{1}\!3 \cdot {}^{4}\!8}{{}_{1}2 \cdot {}_{1}3} = \dfrac{1 \cdot 4}{1 \cdot 1} = \dfrac{4}{1} = 4$

Example 7: $\quad \dfrac{3}{4} \div \dfrac{7}{8} = \dfrac{3}{4} \cdot \dfrac{8}{7} = \dfrac{3 \cdot {}^{2}\!8}{{}_{1}4 \cdot 7} = \dfrac{3 \cdot 2}{1 \cdot 7} = \dfrac{6}{7}$

Example 8: $\quad \dfrac{2}{3} \div \dfrac{1}{6} = \dfrac{2}{3} \cdot \dfrac{6}{1} = \dfrac{2 \cdot {}^{2}\!6}{{}_{1}3 \cdot 1} = \dfrac{2 \cdot 2}{1 \cdot 1} = \dfrac{4}{1} = 4$

Practice problems. Multiply and reduce your answer where necessary.

1. $\frac{2}{5} \times \frac{4}{7}$

2. $\frac{2}{7} \times \frac{2}{9}$

3. $4 \times \frac{3}{8}$ a. 4   b. $4\frac{3}{8}$ c. 2   d. $1\frac{1}{2}$

4. $6 \times \frac{5}{9}$ a. 3   b. $5\frac{5}{9}$ c. $3\frac{4}{9}$ d. $3\frac{1}{3}$

5. $\frac{19}{20} \times \frac{4}{19}$

6. $\frac{8}{9} \times \frac{3}{8}$

7. What is the product of $\frac{2}{5}$ and $\frac{2}{11}$?

8. What is the product of $\frac{2}{3}$ and $\frac{2}{5}$?

9. $\frac{15}{16} \cdot \frac{4}{15}$

10. $\frac{6}{7} \cdot \frac{4}{11}$

11. $14 \times \frac{2}{21}$

12. $6 \times \frac{5}{8}$ a. $3\frac{3}{4}$ b. $6\frac{1}{2}$ c. 4   d. $6\frac{5}{8}$

13. $\frac{9}{10} \cdot \frac{2}{9}$

14. What is the product of $\frac{2}{7}$ and $\frac{2}{3}$?

15. What is the product of $\frac{2}{5}$ and $\frac{4}{11}$?

16. How many $\frac{1}{5}$'s are in $\frac{8}{10}$? $\left(\frac{8}{10} \div \frac{1}{5}\right)$

17. How many $\frac{3}{4}$'s are in $\frac{1}{12}$? $\left(\frac{1}{12} \div \frac{3}{4}\right)$
    a. $\frac{1}{9}$ b. 16   c. $\frac{1}{16}$ d. 9

18. How many $\frac{4}{5}$'s are in $\frac{1}{20}$? $\left(\frac{1}{20} \div \frac{4}{5}\right)$
    a. $\frac{1}{16}$ b. 25   c. 16   d. $\frac{1}{25}$

19. How many $\frac{1}{5}$'s are in $\frac{9}{20}$? $\left(\frac{9}{20} \div \frac{1}{5}\right)$

20. How many $\frac{2}{3}$'s are in $\frac{1}{6}$? $\left(\frac{1}{6} \div \frac{2}{3}\right)$
    a. 4   b. $\frac{1}{4}$ c. 9   d. $\frac{1}{9}$

21. How many $\frac{1}{3}$'s are in $\frac{5}{6}$? $\left(\frac{5}{6} \div \frac{1}{3}\right)$

22. How many $\frac{1}{5}$'s are in $\frac{11}{15}$? $\left(\frac{11}{15} \div \frac{1}{5}\right)$

23. How many $\frac{1}{7}$'s are in $\frac{11}{14}$? $\left(\frac{11}{14} \div \frac{1}{7}\right)$

24. $\frac{3}{8} \div \frac{1}{4}$

25. $\frac{5}{12} \div \frac{5}{4}$

LESSON 19 – Multiple Fractional Factors

Here we have more than two fractions to multiply.

Example 1: $\dfrac{2}{3} \times \dfrac{3}{4} \times \dfrac{15}{60} = \dfrac{{}^{1}2 \times {}^{1}3 \times {}^{1}45}{{}_{1}3 \times {}_{2}4 \times {}_{4}60} = \dfrac{1 \times 1 \times 1}{1 \times 2 \times 4} = \dfrac{1}{8}$

Cancel the 3's, reduce 2 and 4 by 2, finally reduce 15 and 60 by 15. If necessary put 15 and 60 in prime factor form. $15 = 3 \times 5$ and $60 = 2 \times 2 \times 3 \times 5$.

Prime factor form:

$$\frac{2}{3} \times \frac{3}{4} \times \frac{15}{60} = \frac{2}{3} \cdot \frac{3}{2 \times 2} \cdot \frac{3 \times 5}{2 \times 2 \times 3 \times 5} =$$

$$\frac{{}^{1}2}{{}_{1}3} \cdot \frac{{}^{1}3}{{}_{1}2 \times 2} \cdot \frac{{}^{1}(3 \times 5)}{2 \times 2 \times {}_{1}(3 \times 5)} = \frac{1 \times 1 \times 1}{1 \times 2 \times 4} = \frac{1}{8}$$

Example 2: $\dfrac{5}{8} \cdot \dfrac{3}{6} \cdot \dfrac{25}{70} = \dfrac{5}{8} \cdot \dfrac{{}^{1}3}{{}_{2}6} \cdot \dfrac{{}^{5}25}{{}_{14}70} = \dfrac{5 \cdot 1 \cdot 5}{8 \cdot 2 \cdot 14} = \dfrac{25}{224}$

Example 3: $\dfrac{1}{4} \times \dfrac{1}{3} \times \dfrac{1}{2} = \dfrac{1 \times 1 \times 1}{4 \times 3 \times 2} = \dfrac{1}{24}$

Example 4: $\dfrac{5}{16} \cdot \dfrac{1}{2} \cdot \dfrac{8}{10} = \dfrac{5}{2 \cdot 2 \cdot 2 \cdot 2} \cdot \dfrac{1}{2} \cdot \dfrac{2 \cdot 2 \cdot 2}{2 \cdot 5} =$

$$\frac{{}^{1}5}{{}_{1}(2 \cdot 2 \cdot 2) \cdot 2} \cdot \frac{1}{2} \cdot \frac{{}^{1}(2 \cdot 2 \cdot 2)}{2 \cdot {}_{1}5} = \frac{1 \cdot 1 \cdot 1}{2 \cdot 2 \cdot 2} = \frac{1}{8}$$

Example 5: $\dfrac{4}{5} \times \dfrac{3}{8} \times \dfrac{20}{55} = \dfrac{{}^{1}4}{{}_{1}5} \times \dfrac{3}{{}_{12}8} \times \dfrac{{}^{24}20}{55} = \dfrac{1 \times 3 \times 2}{1 \times 1 \times 55} = \dfrac{6}{55}$

Example 6: $\dfrac{2}{3} \cdot \dfrac{4}{7} \cdot \dfrac{28}{36} = \dfrac{2}{3} \cdot \dfrac{{}^{1}4}{{}_{1}7} \cdot \dfrac{{}^{4}28}{{}_{9}36} = \dfrac{2 \cdot 1 \cdot 4}{3 \cdot 1 \cdot 9} = \dfrac{8}{27}$

Practice on Multiple Fractional Factors.

1. $\dfrac{2}{5} \times \dfrac{1}{7} \times \dfrac{14}{105}$ a. $\dfrac{20}{525}$ b. $\dfrac{4}{525}$ c. $\dfrac{4}{15}$ d. $\dfrac{4}{105}$

2. $\dfrac{3}{5} \cdot \dfrac{3}{4} \cdot \dfrac{15}{80}$ a. $\dfrac{135}{40}$ b. $\dfrac{27}{16}$ c. $\dfrac{27}{64}$ d. $\dfrac{27}{320}$

3. $\dfrac{1}{7} \times \dfrac{3}{5} \times \dfrac{20}{175}$ a. $\dfrac{84}{1225}$ b. $\dfrac{12}{1225}$ c. $\dfrac{12}{35}$ d. $\dfrac{12}{175}$

4. $\dfrac{2}{7} \times \dfrac{1}{7} \times \dfrac{21}{196}$ a. $\dfrac{42}{1372}$ b. $\dfrac{3}{686}$ c. $\dfrac{3}{98}$ d. $\dfrac{3}{14}$

5. $\dfrac{2}{5} \times \dfrac{3}{4} \times \dfrac{20}{100}$ a. $\dfrac{3}{10}$ b. $\dfrac{3}{50}$ c. $\dfrac{120}{500}$ d. $\dfrac{6}{5}$

6. $\dfrac{1}{5} \cdot \dfrac{1}{7} \cdot \dfrac{21}{140}$ a. $\dfrac{15}{700}$ b. $\dfrac{3}{700}$ c. $\dfrac{3}{20}$ d. $\dfrac{3}{140}$

7. $\dfrac{3}{7} \times \dfrac{3}{5} \times \dfrac{10}{105}$ a. $\dfrac{126}{735}$ b. $\dfrac{6}{7}$ c. $\dfrac{6}{245}$ d. $\dfrac{6}{35}$

8. $\dfrac{1}{3} \cdot \dfrac{1}{4} \cdot \dfrac{9}{48}$ a. $\dfrac{1}{64}$ b. $\dfrac{3}{16}$ c. $\dfrac{9}{144}$ d. $\dfrac{3}{64}$

9. $\dfrac{1}{7} \times \dfrac{1}{7} \times \dfrac{14}{147}$ a. $\dfrac{2}{1029}$ b. $\dfrac{2}{21}$ c. $\dfrac{2}{147}$ d. $\dfrac{14}{1029}$

10. $\dfrac{2}{7} \times \dfrac{3}{5} \times \dfrac{20}{175}$ a. $\dfrac{24}{1225}$ b. $\dfrac{168}{1225}$ c. $\dfrac{24}{175}$ d. $\dfrac{24}{35}$

11. $\dfrac{2}{3} \cdot \dfrac{3}{4} \cdot \dfrac{6}{36}$ a. $\dfrac{1}{4}$ b. 1 c. $\dfrac{36}{108}$ d. $\dfrac{1}{12}$

12. $\dfrac{3}{5} \times \dfrac{1}{7} \times \dfrac{21}{140}$ a. $\dfrac{9}{140}$ b. $\dfrac{9}{700}$ c. $\dfrac{9}{20}$ d. $\dfrac{45}{700}$

LESSON 20 – Changing a Mixed Number to an Improper Fraction

Example 1:   Change $4\frac{5}{8}$ to an improper fraction.

Step 1:   Multiply the whole number by the denominator of the fraction.

Step 2:   Add the numerator of the fraction to the product of step 1.

Step 3:   Rewrite the fraction by taking the step 2 answer as the new numerator over the original fraction's denominator.

$$4 \times 8 = 32$$

$$32 + 5 = 37$$

$$4\frac{5}{8} = \frac{37}{8}$$

In short: $\dfrac{(4 \times 8) + 5}{8}$

Example 2:   Change $3\frac{2}{5}$ to an improper fraction.

Step 1:   Multiply the whole number by the denominator of the fraction.

Step 2:   Add the numerator of the fraction to the product of step 1.

Step 3:   Rewrite the fraction by taking the step 2 answer as the new numerator over the original fraction's denominator.

$$3 \times 5 = 15$$

$$15 + 2 = 17$$

$$3\frac{2}{5} = \frac{17}{5}$$

In short: $\dfrac{(3 \times 5) + 2}{5}$

Change $17\frac{2}{3}$ to an improper fraction.

Step 1:   Multiply the whole number by the denominator of the fraction.

Step 2:   Add the numerator of the fraction to the product of step 1.

Step 3:   Rewrite the fraction by taking the step 2 answer as the new numerator over the original fraction's denominator.

$$17 \cdot 3 = 51$$

$$51 + 2 = 53$$

$$17\frac{2}{3} = \frac{53}{3}$$

In short: $\dfrac{(17 \cdot 3) + 2}{3}$

Now try these for practice. Write as improper fractions.

1. $4\frac{2}{3}$ a. $\frac{12}{3}$ b. $\frac{8}{3}$ c. 14   d. $\frac{14}{3}$

2. $5\frac{1}{2}$ a. $\frac{10}{2}$ b. $\frac{51}{2}$ c. $\frac{11}{2}$ d. $\frac{5}{2}$

3. $2\frac{4}{5}$

4. $1\frac{1}{2}$

5. $4\frac{1}{3}$ a. $\frac{13}{3}$ b. $\frac{41}{3}$ c. $\frac{12}{3}$ d. $\frac{4}{3}$

6. $2\frac{1}{2}$

7. $2\frac{3}{4}$

8. $3\frac{2}{5}$

9. $3\frac{3}{5}$ a. $\frac{18}{5}$ b. $\frac{15}{5}$ c. $\frac{33}{5}$ d. $\frac{9}{5}$

10. $2\frac{1}{3}$ a. $\frac{6}{3}$ b. $\frac{7}{3}$ c. $\frac{2}{3}$ d. 7

11. $5\frac{3}{4}$ a. $\frac{15}{4}$ b. $\frac{53}{4}$ c. $\frac{20}{4}$ d. $\frac{23}{4}$

12. $4\frac{1}{5}$

13. $7\frac{2}{7}$

14. $8\frac{3}{4}$

15. $4\frac{3}{8}$

16. $6\frac{2}{3}$

17. $5\frac{4}{5}$

18. $3\frac{9}{11}$

LESSON 21 – Multiplying Fractions and/or Whole Numbers Together

Remember to cancel when possible (top to bottom or bottom to top).

Example 1: $6 \times \dfrac{1}{2} = \dfrac{6}{1} \times \dfrac{1}{2} = \dfrac{6 \times 1}{1 \times 2} = \dfrac{6}{2} = 3$

Example 2: $5 \cdot \dfrac{2}{3} = \dfrac{5}{1} \cdot \dfrac{2}{3} = \dfrac{5 \cdot 2}{1 \cdot 3} = \dfrac{10}{3} = 3\frac{1}{3}$

Example 3: $\dfrac{2}{3} \cdot \dfrac{5}{7} = \dfrac{2 \cdot 5}{3 \cdot 7} = \dfrac{10}{21}$

Example 4: $\dfrac{4}{5} \times \dfrac{6}{11} = \dfrac{4 \times 6}{5 \times 11} = \dfrac{24}{55}$

Example 5: $\dfrac{2}{3} \times \dfrac{6}{11} = \dfrac{2 \times \overset{2}{6}}{\underset{1}{3} \times 11} = \dfrac{2 \times 2}{1 \times 11} = \dfrac{4}{11}$

Example 6: $\dfrac{5}{8} \cdot \dfrac{4}{5} \cdot \dfrac{7}{12} = \dfrac{\overset{1}{5} \cdot \overset{1}{4} \cdot 7}{\underset{2}{8} \cdot \underset{1}{5} \cdot 12} = \dfrac{1 \cdot 1 \cdot 7}{2 \cdot 1 \cdot 12} = \dfrac{7}{24}$

Example 7: $6 \cdot \dfrac{4}{3} \cdot \dfrac{7}{8} = \dfrac{6}{1} \cdot \dfrac{4}{3} \cdot \dfrac{7}{8} = \dfrac{2 \cdot 3}{1} \cdot \dfrac{(2 \cdot 2)}{3} \cdot \dfrac{7}{2 \cdot (2 \cdot 2)} =$

$\dfrac{\overset{1}{2} \cdot \overset{1}{3}}{1} \cdot \dfrac{\overset{1}{(2 \cdot 2)}}{\underset{1}{3}} \cdot \dfrac{7}{\underset{1}{2} \cdot \underset{1}{(2 \cdot 2)}} = \dfrac{1 \cdot 1 \cdot 7}{1 \cdot 1 \cdot 7} = \dfrac{7}{1} = 7$

Now try these problems. Multiply and reduce your answer where necessary.

1. $6 \times \dfrac{5}{9}$ a. $6\frac{1}{3}$ b. $6\frac{5}{9}$ c. 3 d. $3\frac{1}{3}$

2. $\dfrac{8}{9} \cdot \dfrac{2}{3}$

3. $\dfrac{7}{8} \times \dfrac{4}{7}$

4. What is the product of $\dfrac{2}{3}$ and $\dfrac{5}{7}$?

5. What is the product of $\frac{2}{5}$ and $\frac{2}{3}$?

6. $\frac{11}{12} \times \frac{4}{11}$

7. $\frac{7}{9} \times \frac{4}{5}$

8. $\frac{3}{5} \cdot \frac{9}{11}$

9. $\frac{2}{3} \times \frac{5}{11}$

10. $6 \times \frac{4}{9}$ a. $6\frac{4}{9}$ b. $4\frac{1}{9}$ c. 3   d. $2\frac{2}{3}$

11. $9 \cdot \frac{1}{6}$ a. $2\frac{1}{2}$ b. 2   c. $1\frac{1}{2}$ d. $9\frac{1}{6}$

12. $\frac{8}{9} \times \frac{3}{8}$

13. What is the product of $\frac{2}{5}$ and $\frac{8}{9}$?

14. $\frac{11}{12} \cdot \frac{3}{11}$

15. What is the product of $\frac{2}{5}$ and $\frac{8}{9}$?

16. What is the product of $\frac{2}{3}$ and $\frac{10}{11}$?

17. What is the product of $\frac{4}{9}$ and $\frac{2}{3}$?

18. What is the product of $\frac{4}{7}$ and $\frac{2}{5}$?

19. What is the product of $\frac{2}{9}$ and $\frac{2}{3}$?

20. $6 \cdot \frac{5}{8}$ a. $3\frac{3}{4}$ b. 4   c. $6\frac{5}{8}$ d. $6\frac{1}{8}$

21. $\frac{5}{7} \times \frac{8}{11}$

22. What is the product of $\frac{4}{5}$ and $\frac{2}{3}$?

23. $\frac{4}{7} \cdot \frac{14}{9}$

24. $8 \times \frac{1}{2}$

25. $\frac{3}{8} \cdot \frac{12}{15}$

26. $\frac{7}{20} \times \frac{15}{35}$

27. $\frac{8}{14} \times \frac{7}{40} \times \frac{2}{3}$

28. $\frac{3}{5} \cdot \frac{15}{21} \cdot \frac{1}{4}$

29. $\frac{5}{8} \times \frac{2}{3} \times \frac{8}{5}$

30. $\frac{2}{3} \cdot \frac{1}{2} \cdot \frac{9}{12}$

31. $16 \cdot \frac{5}{8} \cdot \frac{2}{30}$

32. $\frac{4}{5} \times \frac{7}{8} \times \frac{5}{28}$

33. $\frac{2}{7} \cdot \frac{7}{9} \cdot \frac{3}{2}$

34. $\frac{8}{20} \cdot 24 \cdot \frac{15}{18}$

35. $\frac{4}{5} \cdot \frac{5}{2} \cdot \frac{4}{7}$

LESSON 22 – Dividing Fractions

There are two ways to write division of fractions:

$$\frac{a}{b} \div \frac{c}{d} \quad \text{or} \quad \frac{\frac{a}{b}}{\frac{c}{d}}$$

Take the reciprocal of the second (or bottom) fraction and multiply times the first (or top) fraction.

Reciprocal: Two numbers, $\frac{a}{b}$ and $\frac{b}{a}$, whose product is 1.

$$\frac{a}{b} \div \frac{c}{d} = \frac{a}{b} \times \frac{d}{c} = \frac{a \times d}{b \times c} = \frac{ad}{bc}$$

reciprocal

$$\frac{\frac{a}{b} \times \frac{d}{c}}{\frac{c}{d} \times \frac{d}{c}} = \frac{\frac{a}{b} \times \frac{d}{c}}{1} = \frac{a}{b} \cdot \frac{d}{c} = \frac{a \cdot d}{b \cdot c} = \frac{ad}{bc}$$

Example 1: $\dfrac{4}{5} \div \dfrac{3}{8} = \dfrac{4}{5} \cdot \dfrac{8}{3} = \dfrac{4 \cdot 8}{5 \cdot 3} = \dfrac{32}{15} = 2\dfrac{2}{15}$

Example 2: $\dfrac{5}{8} \div \dfrac{3}{4} = \dfrac{5}{8} \times \dfrac{4}{3} = \dfrac{5 \times \overset{1}{\cancel{4}}}{\underset{2}{\cancel{8}} \times 3} = \dfrac{5 \times 1}{2 \times 3} = \dfrac{5}{6}$

Example 3: $\dfrac{3}{4} \div 8 = \dfrac{3}{4} \div \dfrac{8}{1} = \dfrac{3}{4} \times \dfrac{1}{8} = \dfrac{3 \times 1}{4 \times 8} = \dfrac{3}{32}$

Example 4: $\dfrac{\frac{5}{7}}{\frac{3}{8}} = \dfrac{\frac{5}{7} \cdot \frac{8}{3}}{\frac{3}{8} \cdot \frac{8}{3}} = \dfrac{\frac{40}{21}}{\frac{24}{24}} = \dfrac{\frac{40}{21}}{1} = \dfrac{40}{21} = 1\dfrac{19}{21}$

Example 5: $\dfrac{\frac{9}{16}}{\frac{1}{4}} = \dfrac{\frac{9}{16} \times \frac{4}{1}}{\frac{1}{4} \times \frac{4}{1}} = \dfrac{\frac{36}{16}}{\frac{4}{4}} = \dfrac{\frac{36}{16}}{1} = \dfrac{36}{16} = 2\dfrac{4}{16} = 2\dfrac{1}{4}$

Try these for practice. Divide and reduce your answer where necessary.

1. $\dfrac{2}{3} \div 5$

2. $\dfrac{2}{3} \div \dfrac{1}{5}$

3. $\dfrac{2}{5} \div \dfrac{3}{5}$

4. $10 \div \dfrac{1}{5}$

5. $\dfrac{7}{8} \div \dfrac{3}{8}$

6. $\dfrac{5}{8} \div \dfrac{5}{9}$

7. $\dfrac{3}{4} \div \dfrac{7}{8}$

8. $\dfrac{6}{5} \div 10$

9. $\dfrac{\frac{5}{9}}{\frac{1}{7}}$

10. $\dfrac{\frac{2}{3}}{\frac{3}{8}}$

11. $\dfrac{\frac{5}{16}}{\frac{3}{4}}$

12. $\dfrac{\frac{9}{8}}{\frac{8}{9}}$

13. $\dfrac{\frac{9}{8}}{\frac{9}{8}}$

14. $\dfrac{\frac{4}{7}}{\frac{9}{5}}$

15. $\dfrac{\frac{5}{8}}{\frac{3}{4}}$

16. $\dfrac{\frac{3}{8}}{\frac{5}{16}}$

17. $14 \div \dfrac{2}{3}$

18. $\dfrac{\frac{6}{7}}{\frac{15}{19}}$

19. $\dfrac{\frac{12}{13}}{\frac{2}{5}}$

20. $\dfrac{17}{24} \div \dfrac{5}{8}$

LESSON 23 – Multiplying and Dividing Fractions and Mixed Numbers

Remember: You must change mixed numbers to improper fractions before multiplying and/or dividing.

Example 1: $2\frac{3}{4} \div 1\frac{5}{8} = \frac{11}{4} \div \frac{13}{8} = \frac{11}{4} \cdot \frac{8}{13} = \frac{11}{\cancel{4}_1} \cdot \frac{\cancel{8}^2}{13} = \frac{11 \cdot 2}{1 \cdot 13} = \frac{22}{13} = 1\frac{9}{13}$

Example 2: $\dfrac{4\frac{7}{12}}{3\frac{2}{3}} = \dfrac{\frac{55}{12}}{\frac{11}{3}} = \frac{55}{12} \div \frac{11}{3} = \frac{55}{12} \times \frac{3}{11} = \frac{\cancel{55}^5}{\cancel{12}_4} \times \frac{\cancel{3}^1}{\cancel{11}_1} = \frac{5 \times 1}{4 \times 1} = \frac{5}{4} = 1\frac{1}{4}$

Example 3: $\frac{5}{8} \times \frac{3}{5} \times \frac{4}{9} = \frac{\cancel{5}^1 \times \cancel{3}^1 \times \cancel{4}^1}{\cancel{8}_2 \times \cancel{5}_1 \times \cancel{9}_3} = \frac{1 \times 1 \times 1}{2 \times 1 \times 3} = \frac{1}{6}$

Example 4: $2\frac{1}{4} \times \frac{5}{8} = \frac{9}{4} \times \frac{5}{8} = \frac{9 \times 5}{4 \times 8} = \frac{45}{32} = 1\frac{13}{32}$

Example 5: $4\frac{1}{5} \cdot 3\frac{2}{3} = \frac{21}{5} \cdot \frac{11}{3} = \frac{231}{15} = 15\frac{6}{15} = 15\frac{2}{5}$

or $\frac{21}{5} \cdot \frac{11}{3} = \frac{\cancel{21}^7}{5} \cdot \frac{11}{\cancel{3}_1} = \frac{77}{5} = 15\frac{2}{5}$

Example 6: $4\frac{5}{8} \div 2\frac{1}{3} = \frac{37}{8} \div \frac{7}{3} = \frac{37}{8} \times \frac{3}{7} = \frac{111}{56} = 1\frac{55}{56}$

Example 7: $5\frac{7}{8} \cdot 4\frac{1}{4} = \frac{47}{8} \cdot \frac{17}{4} = \frac{799}{32} = 24\frac{31}{32}$

Now try these. Multiply or divide as indicated. Reduce your answer where necessary.

1. $1\frac{1}{9} \div 2\frac{7}{9}$ a. $\frac{2}{5}$ b. $1\frac{4}{5}$ c. $8\frac{1}{10}$ d. $\frac{11}{18}$

2. $4\frac{1}{6} \div 8\frac{1}{3}$ a. $\frac{1}{2}$ b. $\frac{18}{25}$ c. $2\frac{8}{9}$ d. $\frac{3}{5}$

3. $\frac{6}{7} \times \frac{3}{5} \times \frac{6}{35}$ a. none of these   b. $\frac{108}{35}$ c. $\frac{108}{1225}$ d. $\frac{108}{175}$

4. $\frac{4}{5} \cdot \frac{1}{7} \cdot \frac{1}{8}$ a. $\frac{1}{2}$ b. none of these   c. $\frac{1}{14}$ d. $\frac{1}{70}$

5. $\dfrac{1\frac{2}{3}}{2\frac{1}{5}}$

6. $\dfrac{13\frac{1}{2}}{6\frac{1}{2}}$

7. $2\frac{2}{9} \div 3\frac{4}{7}$

8. $5\frac{1}{6} \cdot 1\frac{1}{5} \cdot 2\frac{1}{2}$

9. $\dfrac{1}{3} \times \dfrac{3}{8} \times \dfrac{1}{6}$

10. $\dfrac{2}{3} \times \dfrac{3}{5} \times \dfrac{3}{4}$

11. $6\frac{6}{7} \cdot 1\frac{4}{7} \cdot 3\frac{1}{2}$

12. $6\frac{1}{2} \times 3\frac{5}{6} \times 4\frac{6}{7}$

13. $1\frac{2}{3} \times 5\frac{2}{3} \times 3\frac{1}{3}$

14. $2\frac{2}{5} \cdot 5\frac{1}{2} \cdot 3\frac{1}{4}$

15. $\dfrac{3\frac{1}{2}}{4\frac{2}{3}}$

16. $\dfrac{1\frac{4}{7}}{4\frac{3}{4}}$

17. $\dfrac{6}{7} \times \dfrac{7}{5} \times \dfrac{6}{35}$ a. $\frac{36}{35}$ b. $\frac{36}{175}$ c. $\frac{252}{1715}$ d. none of these

18. $6\frac{2}{5} \div 1\frac{7}{9}$ a. $2\frac{1}{4}$ b. $1\frac{13}{32}$ c. $3\frac{3}{5}$ d. $2\frac{5}{14}$

LESSON 24 – Adding Mixed Numbers

Example 1: $6\frac{2}{3} + 4\frac{3}{4}$

Step 1:   Convert the fractions $\frac{2}{3}$ and $\frac{3}{4}$ to the same denominator. Get the LCM for 3 and 4, which is 12.

$$6\frac{2}{3} = 6\frac{8}{12}$$
$$+\ 4\frac{3}{4} = 4\frac{9}{12}$$

$$\begin{array}{ccc} \frac{2}{3} & = & \frac{x}{12} \\ 3x & = & 24 \\ x & = & 8 \\ \frac{2}{3} & = & \frac{8}{12} \end{array}$$

Step 2:   Add the whole numbers and the fractions.

$$6\frac{2}{3} = 6\frac{8}{12}$$
$$+\ 4\frac{3}{4} = 4\frac{9}{12}$$
$$\overline{\qquad\qquad 10\frac{17}{12}} \bigg\} \text{OOPS! An improper fraction!}$$

$$\begin{array}{ccc} \frac{3}{4} & = & \frac{x}{12} \\ 4x & = & 36 \\ x & = & 9 \\ \frac{3}{4} & = & \frac{9}{12} \end{array}$$

Step 3:   Simplify the improper fraction and add to the whole number.

$$10\frac{17}{12} = 10 + \frac{17}{12} = 10 + 1\frac{5}{12} = 11\frac{5}{12}$$

Example 2: $4\frac{2}{5} + 5\frac{3}{10}$

$$4\frac{2}{5} = 4\frac{4}{10}$$
$$+\ 5\frac{3}{10} = 5\frac{3}{10}$$
$$\overline{\qquad\qquad 9\frac{7}{10}}$$

Example 3: $5\frac{7}{8} + 4\frac{2}{3}$

$$5\frac{7}{8} = 5\frac{21}{24}$$
$$+\ 4\frac{2}{3} = 4\frac{16}{24}$$
$$\overline{\qquad\qquad 9\frac{37}{24} = 10\frac{13}{24}}$$

Example 4: $2\frac{1}{2} + 3\frac{2}{3}$

$$2\frac{1}{2} = 2\frac{3}{6}$$
$$+\ 3\frac{2}{3} = 3\frac{4}{6}$$
$$\overline{\qquad\qquad 5\frac{7}{6} = 6\frac{1}{6}}$$

Example 5: $8\frac{4}{5} + 6\frac{2}{3}$

$$8\frac{4}{5} = 8\frac{12}{15}$$
$$+\ 6\frac{2}{3} = 6\frac{10}{15}$$
$$\overline{\qquad\qquad 14\frac{22}{15} = 15\frac{7}{15}}$$

Now try these. Remember to reduce your answer where necessary.

1. $4\frac{2}{3} + 8\frac{3}{5}$

2. $9\frac{1}{3} + 3\frac{3}{10}$

3. $6\frac{2}{3} + 5\frac{3}{4}$

4. $2\frac{1}{3} + 1\frac{3}{8}$

5. $7\frac{1}{3} + 8\frac{3}{5}$

6. $5\frac{1}{5} + 2\frac{1}{2}$ a. $7\frac{2}{7}$ b. 8 c. $7\frac{7}{10}$ d. $5\frac{1}{7}$

7. $7\frac{9}{10} + 1\frac{1}{9}$ a. $8\frac{10}{19}$ b. $8\frac{17}{19}$ c. $9\frac{1}{90}$ d. 9

8. $8\frac{3}{7} + 1\frac{4}{5}$ a. $10\frac{8}{35}$ b. 10 c. $9\frac{7}{12}$ d. $8\frac{2}{3}$

9. $5\frac{2}{3} + 9\frac{3}{4}$

10. $2\frac{2}{3} + 4\frac{3}{5}$

11. $5\frac{1}{5} + 1\frac{1}{6}$ a. $5\frac{7}{8}$ b. 7 c. $6\frac{11}{30}$ d. $6\frac{3}{16}$

12. $2\frac{2}{9} + 1\frac{3}{7}$ a. $5\frac{5}{8}$ b. 4 c. $3\frac{5}{16}$ d. $3\frac{41}{63}$

13. $3\frac{3}{4} + 5\frac{2}{5}$

14. $6\frac{3}{8} + 4\frac{3}{5}$

15. $7\frac{2}{3} + 9\frac{4}{5}$

16. $4\frac{5}{7} + 2\frac{1}{2}$

17. $8\frac{1}{5} + 2\frac{2}{3} + 1\frac{3}{10}$

18. $4\frac{1}{5} + 5\frac{3}{7} + 6\frac{11}{35}$

LESSON 25 – Adding Fractions and Mixed Numbers

Add one more fraction or mixed number to the group.

Example 1: $\dfrac{2}{3} + \dfrac{1}{2} + \dfrac{3}{4}$

LCM (2, 3, 4) = 12

$$\dfrac{2}{3} = \dfrac{8}{12}$$

$$\dfrac{1}{2} = \dfrac{6}{12}$$

$$\dfrac{3}{4} = \dfrac{9}{12}$$

$$\dfrac{23}{12} = 1\dfrac{11}{12}$$

Example 2: $3\dfrac{2}{5} + 4\dfrac{1}{2} + 5\dfrac{3}{4}$

LCM (2, 4, 5) = 20

$$3\dfrac{2}{5} = 3\dfrac{8}{20}$$

$$4\dfrac{1}{2} = 4\dfrac{10}{20}$$

$$5\dfrac{3}{4} = 5\dfrac{15}{20}$$

$$12\dfrac{33}{20} = 13\dfrac{13}{20}$$

Example 3: $4\dfrac{2}{3} + 5\dfrac{1}{2} + 2\dfrac{3}{5}$

LCM (2, 3, 5) = 30

$$4\dfrac{2}{3} = 4\dfrac{20}{30}$$

$$5\dfrac{1}{2} = 5\dfrac{15}{30}$$

$$2\dfrac{3}{5} = 2\dfrac{18}{30}$$

$$11\dfrac{53}{30} = 12\dfrac{23}{30}$$

Try these for extra practice. Remember to reduce your answer where necessary.

1. $5\frac{3}{5} + 6\frac{3}{4} + 6\frac{1}{2}$

2. $2\frac{1}{2} + 5\frac{2}{9} + 4\frac{2}{3}$

3. $\frac{4}{5} + \frac{7}{12} + \frac{1}{6}$ a. none of these b. $2\frac{13}{15}$ c. $1\frac{11}{20}$ d. $\frac{43}{60}$

4. $\frac{3}{7} + \frac{21}{35} + \frac{1}{10}$ a. none of these b. $1\frac{9}{70}$ c. $\frac{79}{140}$ d. $\frac{25}{52}$

5. $\frac{1}{5} + \frac{7}{20} + \frac{1}{8}$ a. $1\frac{7}{20}$ b. $\frac{3}{11}$ c. none of these d. $\frac{27}{80}$

6. $\frac{1}{3} + \frac{7}{15} + \frac{1}{10}$ a. $\frac{9}{10}$ b. $\frac{9}{28}$ c. none of these d. $1\frac{4}{5}$

7. $2\frac{1}{5} + 9\frac{3}{4} + 6\frac{1}{2}$

8. $5\frac{1}{2} + 8\frac{1}{9} + 3\frac{2}{3}$

9. $8\frac{1}{5} + 3\frac{2}{9} + 7\frac{2}{3}$

10. $7\frac{1}{3} + 7\frac{7}{16} + 9\frac{1}{4}$

11. $6\frac{1}{2} + 4\frac{1}{9} + 5\frac{2}{3}$

12. $\frac{2}{7} + \frac{19}{21} + \frac{1}{6}$

13. $\frac{3}{5} + \frac{13}{20} + \frac{1}{8}$

14. $4\frac{2}{3} + 8\frac{1}{4} + 3\frac{1}{2}$

15. $2\frac{6}{7} + 4\frac{4}{9} + 2\frac{1}{3}$

16. $6\frac{1}{2} + 5\frac{4}{9} + 5\frac{2}{3}$

17. $\frac{2}{3} + \frac{1}{15} + \frac{1}{10}$ a. $\frac{5}{6}$ b. $\frac{5}{12}$ c. $1\frac{2}{3}$ d. none of these

18. $\frac{4}{7} + \frac{11}{21} + \frac{1}{6}$ a. none of these   b. $2\frac{11}{21}$ c. $\frac{53}{84}$ d. $\frac{8}{17}$

19. $\frac{2}{7} + \frac{3}{14} + \frac{1}{4}$ a. none of these   b. $\frac{3}{4}$ c. $1\frac{1}{2}$ d. $\frac{21}{25}$

20. $6\frac{1}{7} + 2\frac{5}{9} + 4\frac{1}{3}$

21. $\frac{6}{7} + \frac{14}{35} + \frac{1}{10}$

22. $\frac{2}{3} + \frac{3}{6} + \frac{1}{4}$

23. $2\frac{4}{5} + 5\frac{1}{4} + 7\frac{2}{3}$

24. $2\frac{3}{5} + 17\frac{1}{2} + 12\frac{4}{9}$

25. $4\frac{3}{4} + 7\frac{1}{2} + 9\frac{2}{3}$

26. $3\frac{1}{8} + 5\frac{2}{5} + 6\frac{3}{10}$

27. $6\frac{2}{3} + 7\frac{5}{8} + 8\frac{3}{5}$

28. $2\frac{4}{5} + 4\frac{1}{8} + 6\frac{3}{4}$

29. $3\frac{7}{10} + 5\frac{2}{5} + 7\frac{3}{8}$

30. $17\frac{9}{10} + 19\frac{3}{4} + 21\frac{2}{5}$

31. $15\frac{1}{2} + 13\frac{2}{3} + 10\frac{5}{6}$

32. $22\frac{1}{3} + 13\frac{3}{8} + 4\frac{1}{6}$

LESSON 26 – Subtracting Mixed Numbers

Example 1:
$$4\tfrac{2}{3} = 4\tfrac{4}{6}$$
$$-\ 3\tfrac{1}{2} = 3\tfrac{3}{6}$$
$$\overline{\qquad\quad 1\tfrac{1}{6}}$$

Example 2:
$$15\tfrac{1}{2} = 15\tfrac{2}{4}$$
$$-\ 11\tfrac{1}{4} = 11\tfrac{1}{4}$$
$$\overline{\qquad\quad 4\tfrac{1}{4}}$$

Example 3:
$$8\tfrac{1}{4} = 8\tfrac{3}{12} = 7\tfrac{15}{12} \longleftarrow$$
$$-\ 5\tfrac{2}{3} = 5\tfrac{8}{12} = 5\tfrac{8}{12}$$
$$\overline{\qquad\qquad\qquad\quad 2\tfrac{7}{12}}$$

Cannot subtract 8 from 3. Need to borrow.

$$8\tfrac{3}{12} = 8 + \tfrac{3}{12} = 7 + 1 + \tfrac{3}{12} = 7 + \tfrac{12}{12} + \tfrac{3}{12} = 7 + \tfrac{15}{12} = 7\tfrac{15}{12}$$

The shortcut is to borrow 1 from the whole number. Add the numerator and denominator together and place the sum over the original denominator.

$$8\tfrac{3}{12} = 7\tfrac{3+12}{12} = 7\tfrac{15}{12}$$

Example 4:
$$12\tfrac{1}{5} = 12\tfrac{4}{20} = 11\tfrac{24}{20}$$
$$-\ 7\tfrac{1}{4} = 7\tfrac{5}{20} = 7\tfrac{5}{20}$$
$$\overline{\qquad\qquad\qquad\quad 4\tfrac{19}{20}}$$

Example 5:
$$16\tfrac{3}{8} = 16\tfrac{15}{40} = 15\tfrac{55}{40}$$
$$-\ 7\tfrac{4}{5} = 7\tfrac{32}{40} = 7\tfrac{32}{40}$$
$$\overline{\qquad\qquad\qquad\quad 8\tfrac{23}{40}}$$

Now try these.

1. $14\frac{1}{6} - 1\frac{7}{8}$ a. $12\frac{7}{24}$ b. 13   c. $\frac{11}{686}$ d. $12\frac{17}{48}$

2. $11 - 1\frac{2}{3}$ a. $9\frac{3}{7}$ b. $\frac{8}{199}$ c. $9\frac{1}{3}$ d. 10

3. $8\frac{1}{2} - 1\frac{2}{3}$ a. $6\frac{1}{2}$ b. $6\frac{5}{6}$ c. 7   d. $\frac{5}{77}$

4. $4\frac{1}{4} - 1\frac{8}{9}$ a. 3   b. $2\frac{17}{36}$ c. $\frac{1}{193}$ d. $2\frac{13}{36}$

5. $9\frac{1}{2} - 1\frac{4}{5}$ a. $7\frac{7}{10}$ b. $7\frac{4}{5}$ c. $\frac{6}{77}$ d. 8

6. $15\frac{1}{8} - 1\frac{8}{9}$ a. $13\frac{17}{72}$ b. 14   c. $\frac{12}{1097}$ d. $13\frac{7}{36}$

7. $7\frac{1}{3} - 1\frac{3}{4}$

8. $8\frac{1}{7} - 1\frac{7}{8}$

9. $15\frac{1}{9} - 1\frac{4}{5}$

10. $6\frac{1}{6} - 1\frac{3}{4}$

11. $9\frac{1}{8} - 1\frac{5}{6}$

12. $13\frac{1}{5} - 1\frac{8}{9}$

13. $5\frac{1}{5} - 3\frac{2}{3}$

14. $9\frac{1}{3} - 4\frac{3}{4}$

15. $23\frac{3}{10} - 15\frac{2}{3}$

16. $16\frac{3}{5} - 14\frac{7}{8}$

17. $16\frac{3}{5} - 12$

18. $33 - 18\frac{7}{9}$

LESSON 27 – Adding and Subtracting Fractions

Some review of both adding and subtracting fractions.

1. Subtract: $2\frac{2}{5} - 1\frac{3}{7}$ a. $1\frac{1}{35}$ b. $\frac{34}{35}$ c. $1\frac{1}{7}$ d. $\frac{24}{35}$

2. Add: $3\frac{3}{5} + 1\frac{3}{7}$ a. $5\frac{1}{35}$ b. $4\frac{1}{2}$ c. 5 d. $7\frac{1}{3}$

3. $3\frac{3}{4} + 2\frac{11}{12} - 2\frac{1}{3}$

4. $4\frac{1}{6} + 4\frac{1}{12} - 3\frac{2}{3}$

5. Subtract: $2 - 1\frac{5}{9}$

6. Subtract: $4\frac{3}{8} - 2\frac{1}{3}$

7. Add: $6\frac{4}{5} + 8\frac{3}{7}$

8. Add: $8\frac{3}{4} + 5\frac{3}{10}$

9. Subtract: $6\frac{1}{5} - 3\frac{5}{6}$

10. Subtract: $7\frac{1}{7} - 1\frac{1}{4}$

11. $5\frac{3}{4} + 3\frac{1}{12} - 3\frac{1}{6}$

12. $2\frac{1}{6} + 4\frac{11}{12} - 2\frac{3}{4}$

13. Add: $3\frac{1}{2} + 2\frac{2}{3}$ a. $5\frac{3}{7}$ b. $5\frac{4}{7}$ c. $6\frac{1}{6}$ d. 6

14. Add: $7\frac{1}{4} + 1\frac{1}{8}$ a. $8\frac{1}{8}$ b. $8\frac{3}{8}$ c. $8\frac{3}{16}$ d. 9

15. Add: $3\frac{2}{5} + 1\frac{1}{9}$ a. $4\frac{23}{45}$ b. $7\frac{9}{14}$ c. 5 d. $4\frac{3}{14}$

16. Add: $7\frac{1}{9} + 1\frac{1}{2}$ a. $8\frac{3}{13}$ b. $6\frac{10}{13}$ c. 9 d. $8\frac{11}{18}$

17. Subtract: $4\frac{2}{3} - 3\frac{7}{8}$ a. $1\frac{11}{48}$ b. $\frac{19}{24}$ c. $\frac{7}{48}$ d. $1\frac{1}{16}$

18. Subtract: $2\frac{3}{4} - 1\frac{9}{10}$ a. $1\frac{3}{80}$ b. $\frac{49}{80}$ c. $1\frac{3}{16}$ d. $\frac{17}{20}$

19. Subtract: $3\frac{1}{4} - 1\frac{1}{3}$ a. $2\frac{1}{72}$ b. $1\frac{3}{4}$ c. $1\frac{11}{12}$ d. $2\frac{5}{72}$

20. $2\frac{1}{3} + 4\frac{1}{12} - 4\frac{1}{6}$

21. $17 - 8\frac{4}{5}$

22. $(16\frac{2}{3} + 14\frac{1}{9}) - 3\frac{2}{5}$

23. $25\frac{2}{3} - 15\frac{3}{4}$

24. $7\frac{3}{5} + 8\frac{2}{3} + 9\frac{1}{2}$

25. $(8 + 7\frac{2}{3}) - 14\frac{4}{5}$

26. $(25\frac{2}{5} - 4\frac{3}{5}) + 6$

27. $9\frac{1}{8} + 8\frac{1}{7} + 7\frac{1}{4}$

28. $(5\frac{3}{8} - 2\frac{7}{8}) + 4\frac{4}{9}$

29. $(12\frac{3}{4} + 16\frac{2}{3}) - 27\frac{1}{6}$

30. $(8\frac{2}{3} + 8\frac{2}{3}) - 7\frac{2}{5}$

31. $(16\frac{3}{5} - 8\frac{2}{3}) + 9\frac{3}{8}$

32. $(5\frac{1}{2} + 3\frac{2}{3}) - (4 + 1\frac{7}{8})$

33. $(14\frac{2}{3} - 8\frac{1}{2}) + 5\frac{1}{5}$

34. $(12\frac{3}{4} - 8\frac{1}{3}) + 7\frac{1}{5}$

35. $(9\frac{3}{8} + 17\frac{5}{6}) - 16\frac{5}{8}$

LESSON 28 – More Adding and Subtracting of Mixed Numbers
Removal of Parentheses Before Solving

Example 1: $(6\frac{2}{3} + 4\frac{1}{5}) - 5\frac{1}{3}$

Work inside the parentheses first.

$$
\begin{array}{rcl}
6\frac{2}{3} & = & 6\frac{10}{15} \\
+\,4\frac{1}{5} & = & +\,4\frac{3}{15} \\
\hline
10\frac{13}{15} & = & 10\frac{13}{15} \\
-\,5\frac{1}{3} & = & -\,5\frac{5}{15} \\
\hline
 & & 5\frac{8}{15}
\end{array}
$$

Example 2: $(5\frac{3}{8} + 4\frac{1}{4}) - 3\frac{11}{12}$

$$
\begin{array}{rcccl}
5\frac{3}{8} & = & 5\frac{3}{8} & & \\
+\,4\frac{1}{4} & = & +\,4\frac{2}{8} & & \\
\hline
9\frac{5}{8} & = & 9\frac{15}{24} & = & 8\frac{39}{24} \\
-\,3\frac{11}{12} & = & -\,3\frac{22}{24} & = & -\,3\frac{22}{24} \\
\hline
 & & & & 5\frac{17}{24}
\end{array}
$$

Example 3: $(5\frac{1}{6} - 2\frac{1}{2}) - 1\frac{3}{4}$

$$
\begin{array}{rcccl}
5\frac{1}{6} & = & 5\frac{1}{6} & = & 4\frac{7}{6} \\
-\,2\frac{1}{2} & = & -\,2\frac{3}{6} & = & -\,2\frac{3}{6} \\
\hline
2\frac{4}{6} & = & 2\frac{8}{12} & = & 1\frac{20}{12} \\
-\,1\frac{3}{4} & = & -\,1\frac{9}{12} & = & -\,1\frac{9}{12} \\
\hline
 & & & & \frac{11}{12}
\end{array}
$$

Example 4: $(7\frac{4}{9} - 3\frac{5}{8}) + 2\frac{2}{3}$

$$
\begin{array}{ccccc}
7\frac{4}{9} & = & 7\frac{32}{72} & = & 6\frac{104}{72} \\
-3\frac{5}{8} & = & -3\frac{45}{72} & = & -3\frac{45}{72} \\
\hline
& & & & 3\frac{59}{72} = 3\frac{59}{72} \\
& & & & +2\frac{2}{3} = +2\frac{48}{72} \\
\hline
& & & & 5\frac{107}{72} = 6\frac{35}{72}
\end{array}
$$

Try these:

1. $(3\frac{1}{5} + 5\frac{9}{20}) - 5\frac{3}{4}$

2. $(2\frac{1}{4} + 4\frac{1}{12}) - 4\frac{2}{3}$

3. $(4\frac{1}{4} - 2\frac{1}{12}) + 2\frac{5}{6}$

4. $(5\frac{1}{4} + 3\frac{1}{8}) - 3\frac{1}{2}$

5. $(4\frac{1}{5} - 3\frac{11}{30}) + 4\frac{5}{6}$

6. $(5\frac{2}{3} + 2\frac{1}{12}) - 5\frac{5}{6}$

7. $(3\frac{1}{4} + 5\frac{1}{20}) - 3\frac{3}{5}$

8. $(2\frac{1}{6} + 4\frac{7}{30}) + 2\frac{3}{5}$

9. $(4\frac{1}{3} - 2\frac{1}{12}) + 3\frac{3}{4}$

10. $(3\frac{1}{5} + 5\frac{1}{10}) - 4\frac{1}{2}$

11. $(4\frac{2}{3} + 4\frac{1}{15}) - 5\frac{4}{5}$

12. $(8\frac{2}{5} - 3\frac{1}{2}) - 4\frac{3}{4}$

13. $(4\frac{1}{4} + 4\frac{1}{12}) - 3\frac{2}{3}$

14. $(5\frac{2}{5} + 5\frac{3}{20}) - 2\frac{3}{4}$

15. $(3\frac{1}{4} + 3\frac{7}{20}) - 4\frac{4}{5}$

16. $(3\frac{2}{5} - 2\frac{1}{15}) + 3\frac{2}{3}$

17. $(4\frac{1}{6} - 3\frac{5}{12}) + 2\frac{3}{4}$

18. $(5\frac{1}{4} + 4\frac{1}{8}) - 4\frac{1}{2}$

19. $(5\frac{1}{8} - 2\frac{1}{6}) - 1\frac{5}{12}$

20. $(2\frac{1}{6} + 4\frac{17}{30}) - 4\frac{4}{5}$

LESSON 29 – Percents

Percent, by definition, is a ratio of a number to 100. The symbol for percent is: %.

Example 1:   Express $\frac{24}{100}$ as a percent.

$$\frac{24}{100} = 24\%$$

Example 2:   Express $\frac{3}{5}$ as a percent.

Step 1: Setup a proportion: $\frac{3}{5} = \frac{n}{100}$

Step 2: Solve the proportion.

$$\frac{3}{5} = \frac{n}{100}$$

$5 \cdot n = 3 \cdot 100$
$5n = 300$
$\frac{5n}{5} = \frac{300}{5}$
$n = 60$

Answer: 60%

Example 3:   Express 20% as a fraction in lowest terms.

$$20\% = \frac{20}{100} = \frac{1 \cdot {}^1\cancel{20}}{5 \cdot {}_1\cancel{20}} = \frac{1}{5}$$

Example 4:   Express 150% as a mixed number in lowest terms.

$$\frac{150}{100} = \frac{3 \cdot {}^1\cancel{50}}{2 \cdot {}_1\cancel{50}} = \frac{3}{2} = 1\frac{1}{2}$$

Example 5:   If 65% of the peaches are ripe, what percent are not ripe?

$$100\% - 65\% = 35\%$$

Practice on percent.

Express as a percent.

1. $\frac{1}{4}$

2. $\frac{5}{8}$

3. $\frac{17}{25}$

4. $\frac{53}{50}$

5. $2\frac{1}{4}$

6. $5\frac{11}{22}$

7. $\frac{3}{10}$

8. $\frac{17}{20}$

9. $4\frac{1}{5}$

10. $\frac{27}{8}$

11. $\frac{12}{25}$

12. $3\frac{3}{4}$

13. $\frac{103}{25}$

14. $12\frac{1}{2}$

15. $\frac{3}{5}$

16. $\frac{3}{4}$

Express as a fraction in lowest terms or a mixed number in simplest form.

17. 72%

18. 165%

19. 14%

20. 40%

21. 175%

22. 19%

23. $5\frac{1}{2}\%$

24. 450%

25. 60%

26. 38%

27. 94%

28. 132%

29. $83\frac{1}{3}\%$

30. 220%

31. 55%

32. 86%

33. An airplane has 65% of its seats filled on this flight. What percent are empty?

34. If the team lost 5 of 20 games, what percent did they win?

35. How many games did the football team win if they lost 20% of 15 games?

36. If 200 people attended a conference and 60% were men, how many women attended the conference?

LESSON 30 – Percent Computations

Example 1:   5 is what percent of 20?

$$\frac{5}{20} = \frac{n}{100}$$
$$20 \cdot n = 5 \cdot 100$$
$$20n = 500$$
$$\frac{20n}{20} = \frac{500}{20}$$
$$n = 25$$

Answer: 25%

Example 2:   8 is what percent of 40?

$$\frac{8}{40} = \frac{n}{100}$$
$$40 \cdot n = 8 \cdot 100$$
$$40n = 800$$
$$\frac{40n}{40} = \frac{800}{40}$$
$$n = 20$$

Answer: 20%

Example 3:   What percent of 60 is 15?

$$\frac{15}{60} = \frac{n}{100}$$
$$60 \cdot n = 15 \cdot 100$$
$$60n = 1500$$
$$\frac{60n}{60} = \frac{1500}{60}$$
$$n = 25$$

Answer: 25%

Practice on percent computations.

1.   6 is what percent of 60?

    a. 0.1%   b. 10%   c. 6%   d. $\frac{1}{10}$%

2. 4 is what percent of 40?

3. 6 is what percent of 30?

4. 9 is what percent of 90?

5. 2 is what percent of 10?

6. 3 is what percent of 30?

7. 5 is what percent of 25?
    a. $\frac{1}{20}$%   b. 0.2%   c. 5%   d. 20%

8. 1 is what percent of 10?
    a. 1%   b. 0.1%   c. 10%   d. $\frac{1}{10}$%

9. 4 is what percent of 40?
    a. $\frac{1}{10}$%   b. 4%   c. 10%   d. 0.1%

10. 8 is what percent of 40?
    a. 20%   b. 0.2%   c. 8%   d. $\frac{1}{20}$%

11. 7 is what percent of 70?
    a. 10%   b. 7%   c. $\frac{1}{10}$%   d. 0.1%

12. 3 is what percent of 15?

13. What percent of 40 is 14?

14. What percent of 90 is 30?

15. What percent of 25 is 14?

16. What percent of 20 is 25?

17. What percent of 50 is 33?

18. What percent of 80 is 16?

LESSON 31 – Solving a Proportion

Example 1: $\dfrac{n}{4} = \dfrac{5}{10}$

Cross multiply, then divide by the coefficient of the variable.

$\dfrac{n}{4} = \dfrac{5}{10}$

$10 \cdot n = 4 \cdot 5$

$10n = 20$

$\dfrac{10n}{10} = \dfrac{20}{10}$

$n = 2$

Example 2: $\dfrac{a}{10} = \dfrac{6}{20}$

$20a = 60$

$a = 3$

Example 3: $\dfrac{b}{5} = \dfrac{3}{10}$

$10b = 15$

$b = \dfrac{3}{2} = 1\dfrac{1}{2}$

Example 4: $\dfrac{a}{b} = \dfrac{c}{d}$

$ad = bc$ REMEMBER!!

Example 5: Is $\dfrac{5}{15} = \dfrac{6}{18}$ a proportion? Yes or No.

According to example 4, ad = bc. Therefore, $5 \cdot 18$ must equal $15 \cdot 6$ in order for this to be classified as a proportion.

$5 \cdot 18 \overset{?}{=} 15 \cdot 6$

$90 \overset{\checkmark}{=} 90$

Yes, $\dfrac{5}{15} = \dfrac{6}{18}$ is a proportion.

Example 6: Is $\dfrac{3}{7} = \dfrac{9}{20}$ a proportion? Yes or No.

$$3 \cdot 20 \overset{?}{=} 7 \cdot 9$$
$$60 \neq 63$$

$\dfrac{3}{7} = \dfrac{9}{20}$ is **not** a proportion.

Practice on proportions.

Solve the proportion.

1. $\dfrac{q}{2} = \dfrac{4}{16}$

2. $\dfrac{g}{5} = \dfrac{2}{50}$

3. $\dfrac{r}{4} = \dfrac{5}{80}$

4. $\dfrac{4}{5} = \dfrac{20}{x}$ a. x = 25   b. x = 10   c. x = 18   d. x = 34

5. $\dfrac{2}{6} = \dfrac{x}{18}$ a. x = 3   b. x = 6   c. x = 11   d. x = 5

6. $\dfrac{3}{x} = \dfrac{6}{14}$ a. x = 7   b. x = 4   c. x = 12   d. x = 10

7. $\dfrac{d}{5} = \dfrac{4}{100}$

8. $\dfrac{2}{4} = \dfrac{x}{12}$ a. x = 5   b. x = 6   c. x = 11   d. x = 3

9. $\dfrac{p}{4} = \dfrac{5}{80}$

10. $\dfrac{4}{5} = \dfrac{x}{15}$ a. x = 5   b. x = 21   c. x = 9   d. x = 12

11. How many of the following are proportions?

$$\frac{3}{4} = \frac{24}{32} \qquad \frac{3}{4} = \frac{18}{24} \qquad \frac{3}{4} = \frac{15}{32} \qquad \frac{3}{4} = \frac{21}{20}$$

12. How many of the following are proportions?

$$\frac{4}{7} = \frac{32}{56} \qquad \frac{4}{7} = \frac{40}{70} \qquad \frac{4}{7} = \frac{28}{49} \qquad \frac{4}{7} = \frac{36}{63}$$

Solve the proportion.

13. $\frac{2}{x} = \frac{10}{35}$ a. x = 6   b. x = 10   c. x = 7   d. x = 12

14. $\frac{t}{5} = \frac{5}{125}$

15. $\frac{f}{4} = \frac{3}{32}$

16. $\frac{g}{2} = \frac{6}{24}$

17. $\frac{p}{3} = \frac{4}{24}$

18. $\frac{2}{5} = \frac{4}{x}$ a. x = 5   b. x = 14   c. x = 8   d. x = 10

19. How many of the following are proportions?

$$\frac{3}{5} = \frac{15}{25} \qquad \frac{3}{5} = \frac{21}{35} \qquad \frac{3}{5} = \frac{18}{30} \qquad \frac{3}{5} = \frac{12}{20}$$

20. How many of the following are proportions?

$$\frac{3}{5} = \frac{9}{15} \qquad \frac{3}{5} = \frac{6}{10} \qquad \frac{3}{5} = \frac{15}{25} \qquad \frac{3}{5} = \frac{12}{10}$$

21. Solve the proportion: $\frac{4}{x} = \frac{12}{21}$

      a. x = 10   b. x = 6   c. x = 4   d. x = 7

LESSON 32 – Converting a Percent to a Decimal

To convert a percent to a decimal, locate the decimal point, move it **two places to the left** and remove the percent sign.

Example 1:   34%

        Step 1:   Locate the decimal point: 34.%
        Step 2:   Move the decimal two places to the left: .34.%
        Step 3:   Remove the percent sign: .34

Example 2:   123% = 123.% = 1.23

Example 3:   65% = 65.% = 0.65

Example 4: $73\frac{1}{2}\%$ = 73.5% = 0.735

Example 5:   450% = 450.% = 4.50 = 4.5   The zero is not needed.

To change a decimal to a percent, move the decimal point **two places to the right** and add the percent sign.

Example 6:   0.45

        Step 1:   Move the decimal two places to the right: 0.45.
        Step 2:   Add the percent sign: 45%

Example 7:   0.69 = 69%

Example 8:   1.34 = 134%

Example 9   0.03 = 3%

Example 10: 0.138 = 13.8%

Example 11: 4.365 = 436.5%

Practice on converting percents to decimals. Write or convert each percents to a decimal numeral.

1. 82%   a. 0.82   b. 82   c. 8.2   d. 0.082

2. 8%

3. 94%

4. 9%

5. 34%   a. 34   b. 3.4   c. 0.34   d. 0.034

6. 46%   a. 0.046   b. 4.6   c. 0.46   d. 46

7. 5%

8. 6%

9. 41%

10. 86%

11. 3%

12. 52%   a. 0.052   b. 0.52   c. 5.2   d. 52

13. 35%

14. 18%

15. 55%

16. 64%

17. 71%   a. 0.071   b. 0.71   c. 71   d. 7.1

18. 89%   a. 0.089   b. 8.9   c. 0.89   d. 89

LESSON 33 – Decimals, Fractions, and Percents

Complete the table.

Fraction	Decimal	Percent
	0.15	
$\frac{7}{10}$		.
		142%
$\frac{12}{25}$		
	0.4	

Solutions

Fraction	Decimal	Percent
$\frac{15}{100} = \frac{3}{20}$	0.15	$0.15 = 15\%$
$\frac{7}{10}$	**0.7**	$0.70 = 70\%$
$1\frac{42}{100}$	$142.\% = 1.42$	142%
$\frac{12}{25}$	$48.\% = 0.48$	$\frac{12}{25} = \frac{n}{100}$ n = 48 48%
$\frac{40}{100} = \frac{2}{5}$	0.4	$0.4 = 0.40 = 40\%$

71

Practice on decimals, fractions, and percents. Complete the chart. Reduce fractions to lowest terms where necessary.

Fraction   Decimal   Percent

1. _____   1.2   _____

2. $\frac{7}{25}$ _____   _____

3. _____   _____   230%

4. $\frac{11}{20}$ _____   _____

5. _____   _____   225%

6. _____   0.45   _____

7. $\frac{1}{10}$ 0.1   _____

   a. 1%   b. none of these   c. 0.1%   d. 10%

8. _____   0.41   41%

   a. $\frac{41}{50}$ b. $\frac{41}{10}$ c. none of these   d. $\frac{41}{100}$

9. $\frac{18}{25}$ _____   72%

   a. 0.72   b. 7.2   c. 72   d. none of these

10. $\frac{17}{25}$ _____   68%

   a. 68   b. none of these   c. 0.68   d. 6.8

11. _____   _____   15%

12. _____   1.7   _____

13. $\frac{1}{5}$ _____   _____

14. _____   _____   180%

15. $\frac{4}{25}$ _____   16%

   a. 16   b. 1.6   c. 0.16   d. none of these

LESSON 34 – Fractional Part of a Number

Example 1: $3\frac{2}{3}$ of what number is 14?

$3\frac{2}{3} \cdot N = 14$ Convert the mixed number to an improper fraction.

$\frac{11}{3} \cdot N = 14$

$\frac{3}{11} \cdot \frac{11}{3} \cdot N = \frac{14}{1} \cdot \frac{3}{11}$ Multiply both sides by the reciprocal.

$N = \frac{42}{11} = 3\frac{9}{11}$

Example 2: $1\frac{2}{5}$ of what number is $9\frac{1}{4}$?

$1\frac{2}{5} \cdot N = 9\frac{1}{4}$

$\frac{7}{5} \cdot N = \frac{37}{4}$

$\frac{5}{7} \cdot \frac{7}{5} \cdot N = \frac{37}{4} \cdot \frac{5}{7}$

$N = \frac{185}{28} = 6\frac{17}{28}$

Practice on fractional part of a number. Simplify answer where necessary.

1. $2\frac{5}{9}$ of what number is 15?

           a. $\frac{5}{69}$ b. $5\frac{20}{23}$ c. $\frac{23}{135}$ d. $13\frac{4}{5}$

2   $3\frac{2}{3}$ of what number is $1\frac{3}{4}$?

3. $1\frac{1}{6}$ of what number is 19?

4. $1\frac{2}{3}$ of what number is 14?

5. $1\frac{8}{15}$ of what number is 19?

6. $1\frac{2}{5}$ of what number is $1\frac{1}{10}$?

7. $2\frac{1}{2}$ of what number is $5\frac{2}{3}$?

8. $2\frac{1}{2}$ of what number is 16?

        a. $6\frac{2}{5}$ b. $1\frac{3}{5}$ c. $\frac{5}{32}$ d. $\frac{5}{8}$

9. $1\frac{7}{10}$ of what number is 14?

        a. $\frac{17}{140}$ b. $8\frac{4}{17}$ c. $\frac{7}{85}$ d. $12\frac{1}{7}$

10. $1\frac{6}{7}$ of what number is $1\frac{1}{4}$?

11. $1\frac{2}{3}$ of what number is 12?

        a. $7\frac{1}{5}$ b. $\frac{4}{5}$ c. $\frac{5}{36}$ d. $1\frac{1}{4}$

12. $1\frac{1}{2}$ of what number is 15?

        a. $\frac{2}{5}$ b. $2\frac{1}{2}$ c. $\frac{1}{10}$ d. 10

13. $1\frac{1}{10}$ of what number is 17?

        a. $15\frac{5}{11}$ b. $6\frac{8}{17}$ c. $\frac{17}{110}$ d. $\frac{11}{170}$

14. $2\frac{1}{3}$ of what number is $1\frac{2}{3}$?

15. $1\frac{3}{10}$ of what number is $5\frac{1}{2}$?

16. $5\frac{1}{2}$ of what number is $1\frac{2}{11}$?

17. $1\frac{1}{2}$ of what number is 13?

18. $1\frac{1}{10}$ of what number is 11?

19. $1\frac{6}{13}$ of what number is 19?

20. $2\frac{3}{5}$ of what number is 14?

LESSON 35 – Decimal Part of a Number

Example 1:   What decimal part of 500 is 125?

$$\frac{125}{500} = \frac{{}^1\cancel{25} \cdot {}^1\cancel{5}}{{}_1\cancel{25} \cdot {}_4\cancel{20}} = \frac{1}{4} = 0.25$$

or

$$\begin{array}{r} 0.25 \\ 500\,\overline{|\,125.00} \\ \underline{100\,0} \\ 25\,00 \\ \underline{25\,00} \\ 0 \end{array}$$

Example 2:   Two-fifths of what number is 90?

$$\frac{2}{5} \cdot N = 90$$

$$\frac{5}{2} \cdot \frac{2}{5} \cdot N = \frac{90}{1} \cdot \frac{5}{2}$$

$$N = \frac{450}{2} = 225$$

Example 3:   What decimal part of 200 is 85?

$$\frac{85}{200} = \frac{{}^1\cancel{5} \cdot 17}{{}_1\cancel{5} \cdot 40} = \frac{17}{40} = 0.425$$

or

$$\begin{array}{r} 0.425 \\ 200\,\overline{|\,85.000} \\ \underline{80\,0} \\ 5\,00 \\ \underline{4\,00} \\ 1\,000 \\ \underline{1\,000} \\ 0 \end{array}$$

Example 4:   Three-quarters of what number is 141?

$$\frac{3}{4} \cdot N = 141$$

$$\frac{4}{3} \cdot \frac{3}{4} \cdot N = \frac{141}{1} \cdot \frac{4}{3}$$

$$N = \frac{564}{3} = 188$$

Practice on decimal part of a number. Simplify answer where necessary.

1.   What decimal part of 880 is 110?

2.   What decimal part of 400 is 300?

3.   Five-tenths of what number is 294?

4.   Eight-tenths of what number is 96?

5.   Seven-tenths of 150 is what number?
      a. 157   b. 105   c. 73.5   d. 103

6.   Nine-tenths of 140 is what number?
      a. 126   b. 149   c. 124   d. 113.4

7.   Three-tenths of 110 is what number?
      a. 9.9   b. 33   c. 113   d. 31

8.   Six-tenths of what number is 192?

9.   Four-tenths of what number is 186?

10.   What decimal part of 1040 is 260?

11.   What decimal part of 360 is 90?

12.   Two-tenths of what number is 258?

13.   Four-tenths of what number is 120?

14.   Three-tenths of 180 is what number?
      a. 56   b. 54   c. 183   d. 16.2

15.   Five-tenths of 50 is what number?
      a. 23   b. 55   c. 12.5   d. 25

76

LESSON 36 – Percent as a Rate

Example 1:   Sixty percent of what number is 120?

Set up a percent proportion: $\dfrac{120}{N} = \dfrac{60}{100}$

Cross multiply: $60 \cdot N = 120 \cdot 100$

$$\dfrac{60}{60} \cdot N = \dfrac{120 \cdot 100}{60}$$

$$N = 200$$

Percent Proportion
$\dfrac{\text{Part}}{\text{What Number}} = \%$

Example 2:   45% of what number is 225?

$$\dfrac{225}{N} = \dfrac{45}{100}$$
$$45N = 22500$$
$$N = 500$$

Example 3:   What percent of 6 is 5?

$$\dfrac{5}{6} = \dfrac{N}{100}$$
$$6N = 500$$
$$N = 83\tfrac{1}{3}\%$$

Example 4:   160% of what number is 200?

$$\dfrac{200}{N} = \dfrac{160}{100}$$
$$160N = 20000$$
$$N = 125$$

Example 5:   125% of 80 is what?

$$\dfrac{N}{80} = \dfrac{125}{100}$$
$$100N = 10000$$
$$N = 100$$

Practice on percent as a rate.

1.   Forty percent of what number is 114?

2.   Twenty-five percent of what number is 67?

3.   Two hundred fifty percent of 90 is what number?

4.   What percent of 20 is 36?

5.   Three hundred seventy percent of what number is 370?

6.   One hundred seventy percent of 90 is what number?

7.   What percent of 14 is 42?

8.   One hundred ninety percent of what number is 570?

9.   What percent of 5.6 is 14?
                a. 40%   b. 250%   c. 265?   d. 35%

10.   What percent of 14.4 is 25.2?
                a. 190%   b. 52%   c. 57%   d. 175%

11.   One hundred seventy percent of 70 is what number?

12.   What percent of 4 is 9?

13.   Two hundred ten percent of what number is 630?

14.   Forty percent of what number is 84?

15.   Twenty-five percent of what number is 46?

16.   Twenty percent of what number is 36?

17.   What percent of 2.4 is 4.2?
                a. 57%   b. 190%   c. 62%   d. 175%

18.   One hundred twenty percent of 20 is what number?

19.   What percent of 18 is 36?

20.   Two hundred sixty percent of what number is 650?

LESSON 37 – More Percents

Example 1: What percent of 300 is 45?

$$\frac{45}{300} = \frac{N}{100}$$
$$300N = 4500$$
$$N = 15\%$$

Percent Proportion
$\dfrac{Part}{What\ Number} = \%$

Example 2: 150 is 30% of what number?

$$\frac{150}{N} = \frac{30}{100}$$
$$30N = 15000$$
$$N = 500$$

Example 3: Find 225% of 72.

$$\frac{P}{72} = \frac{225}{100}$$
$$100P = 16200$$
$$P = 162$$

Example 4: What number is 160% of 150?

$$\frac{N}{150} = \frac{160}{100}$$
$$100N = 24000$$
$$N = 240$$

Example 5: Find 200% of 60.

$$\frac{N}{60} = \frac{200}{100}$$
$$100N = 12000$$
$$N = 120$$

Practice on more percents.

1.   What percent of 1300 is 65?

2.   What percent of 1800 is 36?

3.   What percent of 1600 is 32?

4.   72 is 20% of what number?

5.   78 is 40% of what number?

6.   120 is 20% of what number?

7.   228 is 50% of what number?
                    a. 456   b. 11.4   c. 114   d. 45.6

8.   114 is 60% of what number?
                    a. 190   b. 68.4   c. 684   d. 1900

9.   180 is 20% of what number?
                    a. 900   b. 36   c. 9000   d. 3.6

10.   150 is 40% of what number?

11.   What percent of 1400 is 28?

12.   96 is 20% of what number?

13.   192 is 30% of what number?
                    a. 57.6   b. 640   c. 64   d. 5.76

14.   144 is 40% of what number?
                    a. 3600   b. 576   c. 360   d. 57.5

15.   288 is 20% of what number?

16.   Find 170% of 60.

17.   What number is 140% of 50?
                    a. 70   b. 3571   c. 7000   d. 36

18.   Find 140% of 50.

19.   Find 190% of 30.

20. What number is 120% of 260?
                                    a. 217   b. 21,667   c. 31,200   d. 312

21. Find 140% of 20.

22. Find 170% of 80.

23. What number is 130% of 80?
                                    a. 104   b. 62   c. 10,400   d. 6154

24. What number is 140% of 190?
                                    a. 26,600   b. 136   c. 266   d. 13,571

25. Find 140% of 40.

26. Find 130% of 90.

27. What number is 140% of 270?
                                    a. 193   b. 378   c. 19,286   d. 37,800

28. What number is 130% of 220?
                                    a. 286   b. 169   c. 28,600   d. 16,923

29. Find 160% of 30.

30. Find 130% of 40.

31. 15 is 30% of what number?

32. Five-eighths of what number is 40?

33. Find 175% of 80.

34. 70 is what percent of 200?

35. Find 250% of 60.

36. 160 is 40% of what number?

37. What decimal part of 12 is 7?

38. Seven-fifths of what number is 49?

39. Eighty percent of 72 is what number?

40. What percent of 75 is 250?

LESSON 38 – Percentage Increases

Example 1: What percent of 20 is 32?

$$\frac{32}{20} = \frac{N}{100}$$
20N = 3200
N = 160%

Example 2: It costs $35 to fill up your car with gasoline.

      a.   How much will it cost next year if gas increases by 20%.
      b.   What is the dollar increase?

          a. $35 × 120% = 35 × 1.2 = $42
               or $35 + $35 • 20% =
                       $35 + $35 • 0.2 =
                       $35 + $7 = $42
          b. $42 – $35 = $7

Example 3: Fifteen percent of what number is 45?

$$\frac{45}{N} = \frac{15}{100}$$
15N = 4500
N = 300

Example 4: A basket of tomatoes costs $7.50. The cost of the tomatoes increases by 10%. What is the new price of the tomato basket?

$7.50 × 110% = $7.50 × 1.10 = $8.25
   or $7.50 + $7.50 • 10% =
          $7.50 + $7.50 • 0.1 =
          $7.50 + $0.75 = $8.25

Practice on percentage increases.

1.   What percent of 16 is 44?

2.   What percent of 22 is 44?

3.   Five percent of what number is 340?

4.   Fifty percent of what number is 225?

5.   Ten percent of what number is 200?

6.   What percent of 2 is 4?

7.   The cost of building a house increases 8% every year. If it costs $97,000 to build a house this year, what would it cost to build a house next year?

8.   A department store purchases a dress for $95. To sell the dress to customers, the price is increased by 26%. What is the price of the dress?

9.   Due to a sudden freeze, the cost of apples increased by 30% in one month. If the cost after the increase was 52¢ per pound, what was the cost before the increase?

10.   A department store purchases a coat for $110. To sell the coat to customers, the price is increased by 26%. What is the price of the coat?

11.   The cost of building a house increases 12% every year. If it costs $143,000 to build a house this year, what would it cost to build a house next year?

12.   A department store purchases a coat for $95. To sell the coat to customers, the price is increased by 18%. What is the price of the coat?

13.   The cost of building a house increases 20% every year. If it costs $123,000 to build a house this year, what would it cost to build a house next year?

14.   Bob was earning $6.00 per hour. After working 12 months he received a raise of 20%. What was his hourly rate after the raise?

15.   Due to a truck strike, the cost of bananas increased by 35% in one month. If the cost after the increase was 81¢ per pound, what was the cost before the increase?

LESSON 39 – Solving Proportions with Mixed Numbers

Example 1: $\dfrac{x}{1\frac{1}{2}} = \dfrac{\frac{3}{4}}{2\frac{2}{5}}$

$2\frac{2}{5}x = 1\frac{1}{2} \cdot \frac{3}{4}$ Cross multiply $1\frac{1}{2} \cdot \frac{3}{4} = \frac{3}{2} \cdot \frac{3}{4} = \frac{9}{8}$

$\dfrac{12x}{5} = \dfrac{9}{8}$ Cross multiply again

$96x = 45$ or $\dfrac{5}{12} \cdot \dfrac{12}{5} x = \dfrac{9}{8} \cdot \dfrac{5}{12}$

$x = \dfrac{45}{96} = \dfrac{15}{32}$ $x = \dfrac{^3 9}{8} \cdot \dfrac{5}{_4 12} = \dfrac{3}{8} \cdot \dfrac{5}{4} = \dfrac{15}{32}$

Example 2: $\dfrac{1\frac{3}{4}}{4\frac{1}{2}} = \dfrac{x}{7\frac{1}{3}}$

$4\frac{1}{2}x = 1\frac{3}{4} \cdot 7\frac{1}{3}$ $1\frac{3}{4} \cdot 7\frac{1}{3} = \dfrac{7}{_2 4} \cdot \dfrac{^{11} 22}{3} = \dfrac{77}{6}$

$\dfrac{9x}{2} = \dfrac{77}{6}$

$54x = 154$ or $\dfrac{2}{9} \cdot \dfrac{9}{2} x = \dfrac{77}{6} \cdot \dfrac{2}{9}$

$x = \dfrac{154}{54} = 2\frac{46}{54} = 2\frac{23}{27}$ $x = \dfrac{77}{_3 6} \cdot \dfrac{^1 2}{9} = \dfrac{77}{27} = 2\frac{23}{27}$

Practice on solving proportions with mixed numbers. Simplify answers where necessary.

1. $\dfrac{x}{3\frac{1}{5}} = \dfrac{\frac{1}{4}}{1\frac{1}{2}}$

2. $\dfrac{\frac{1}{5}}{3\frac{1}{2}} = \dfrac{x}{2\frac{6}{7}}$

3. $\dfrac{1\frac{2}{3}}{x} = \dfrac{1\frac{5}{7}}{\frac{3}{5}}$ a. $\dfrac{12}{7}$ b. $\dfrac{100}{21}$ c. $\dfrac{7}{12}$ d. $\dfrac{21}{100}$

84

4. $\dfrac{4\frac{2}{7}}{1\frac{1}{5}} = \dfrac{x}{\frac{2}{3}}$ a. $\dfrac{50}{21}$ b. $\dfrac{54}{7}$ c. $\dfrac{7}{54}$ d. $\dfrac{24}{7}$

5. $\dfrac{\frac{5}{4}}{x} = \dfrac{1\frac{3}{5}}{1\frac{3}{7}}$ a. $\dfrac{125}{112}$ b. $\dfrac{20}{7}$ c. $\dfrac{7}{20}$ d. $\dfrac{64}{35}$

6. $\dfrac{1\frac{1}{5}}{2\frac{2}{3}} = \dfrac{x}{\frac{2}{3}}$ a. $\dfrac{5}{24}$ b. $\dfrac{3}{10}$ c. $\dfrac{24}{5}$ d. $\dfrac{32}{15}$

7. $\dfrac{x}{3\frac{3}{4}} = \dfrac{\frac{2}{3}}{4\frac{2}{3}}$

8. $\dfrac{x}{\frac{3}{4}} = \dfrac{3\frac{3}{7}}{2\frac{1}{4}}$

9. $\dfrac{\frac{1}{3}}{2\frac{1}{2}} = \dfrac{x}{3\frac{3}{5}}$ a. 3   b. $\dfrac{12}{25}$ c. 27   d. $\dfrac{1}{27}$

10. $\dfrac{4\frac{4}{7}}{4\frac{1}{2}} = \dfrac{x}{\frac{1}{2}}$

11. $\dfrac{\frac{1}{3}}{3\frac{1}{4}} = \dfrac{x}{5\frac{2}{5}}$

12. $\dfrac{4\frac{3}{7}}{10\frac{1}{3}} = \dfrac{9\frac{1}{8}}{x}$

13. $\dfrac{6\frac{4}{5}}{8\frac{1}{3}} = \dfrac{x}{12\frac{1}{2}}$

14. $\dfrac{x}{14\frac{2}{7}} = \dfrac{9\frac{1}{3}}{15\frac{7}{9}}$

15. $\dfrac{15\frac{1}{6}}{x} = \dfrac{3\frac{2}{3}}{8\frac{1}{4}}$

LESSON 40 – Easy Story Problems

Example 1: Joe spent $6.18 for 6 pens. How much did each one cost?

$$
\begin{array}{r}
1.03 \\
6\,\overline{)\,6.18} \\
\underline{6} \\
18 \\
\underline{18} \\
0
\end{array}
$$

$1.03 each

Example 2: CD's are on sale: 2 for $9. You bought 7 CDs. How much change did you get from two $20 bills?

$$\frac{7}{2} \times \$9 = \frac{\$63}{2} = \$31.50 \text{ for 7 CDs}$$

2($20) = $40

$40 – $31.50 = $8.50 change

Example 3: Sixteen DVD's cost $132.80. How much did each DVD cost?

$$
\begin{array}{r}
8.30 \\
16\,\overline{)\,132.80} \\
\underline{128} \\
4\;8 \\
\underline{4\;8} \\
0
\end{array}
$$

$8.30 each

Practice problems on easy story problems.

1. A discount audio store advertised a collection of hits from the 1990's on 11 compact discs for $47.52. What is the cost of each disc?

2. A discount audio store advertised a collection of classical music on 14 compact discs for $85.40. What is the cost of each disc?

3. Amber spent $4.27 at the stationery store. If she bought 7 erasers, how much did each eraser cost?
    a. $0.63   b. $0.61   c. $0.60   d. $29.89

4. Adrian spent $2.55 at the stationery. If he bought 5 note pads, how much did each note pad cost?
    a. $0.50   b. $0.53   c. $12.75   d. $0.51

5. A discount audio store advertised a collection of heavy metal music on 24 compact discs for $105.36. What is the cost of each disc?

6. A discount audio store advertised a collection of hits from the 1970's on 13 compact discs for $67.86. What is the cost of each disc?

7. Adrian spent $4.08 at the stationery. If he bought 8 pencils, how much did each pencil cost?

    a. $0.51   b. $12.75   c. $0.50   d. $0.53

8. A discount audio store advertised a collection of hits from the 1960's on 14 compact discs for $79.24. What is the cost of each disc?

9. Adrianna spent $1.24 at the stationery. If she bought 4 pens, how much did each pen cost?

    a. $0.29   b. $0.31   c. $4.96   d. $0.30

10. Patricia paid $595 for 7 nights at a hotel. What was the nightly rate for her room?

    a. $43 per night   b. $170 per night
    c. $85 per night   d. $4165 per night

11. Sam bought a $12.99 DVD and 6 compact discs at 75¢ each. How much change did he receive from a $20 bill?

12. At the flea market, Mary sold seven items which totaled $35.25. The purchaser gave her two twenty dollar bills. How much change did Mary give the purchaser?

13. How much change did Sandi receive after she purchased three CD's at $3.99 each and gave the clerk a $5 and $10 bill?

14. Rumple lost his wallet which had a $10 bill, three $5 bills and seventeen $1 bills. How much did Rumple lose?

15. Four bars of chocolate cost 80¢ each. How many more chocolate bars can Amy purchase if she only has a $5 bill?

16. Shasta purchased three cases of bottled water for $3.50 each. After she gave the clerk a $20 bill, she remembered to get a twelve-pack of Mountain Dew. That item cost $3.75. How much change did she receive?

17. Benny bought six pairs of flip-flops for $47.94 total. What change did he receive from three $20 bill? Could he buy any additional flip-flops with his change?

18. Candy bought three jars of ketchup for $1.20 each along with two dispensers of mustard for $1.50 each. How much did the condiments costs?

LESSON 41 – Ratio Problems

Example 1:   A truck travels 350 miles on 14 gallons of diesel fuel. How many gallons will the truck need to travel 600 miles?

$$\frac{14 \text{ gallons}}{350 \text{ miles}} = \frac{x \text{ gallons}}{600 \text{ miles}} \quad \text{cross multiply}$$

$$350 \text{ miles} \cdot x \text{ gallons} = 14 \text{ gallons} \cdot 600 \text{ miles}$$

$$\frac{{}^1 \cancel{350 \text{ miles}}}{{}_1 \cancel{350 \text{ miles}}} \cdot x \text{ gallons} = \frac{14 \text{ gallons} \cdot 600 \, \cancel{\text{miles}}}{350 \, \cancel{\text{miles}}}$$

$$x \text{ gallons} = \frac{14 \cdot 600}{350} \text{ gallons} = 24 \text{ gallons}$$

Example 2:   A car travels 200 miles on 10 gallons of unleaded gas. How many miles can it travel on 18 gallons?

$$\frac{10 \text{ gallons}}{200 \text{ miles}} = \frac{18 \text{ gallons}}{x \text{ miles}}$$

$$10x = 3600$$

$$x = 360 \text{ miles}$$

Practice problems on ratios.

1.   A bus travels 280 miles on 14 gallons of gas. How many gallons will it need to travel 440 miles?

2.   A van travels 48 miles on 4 gallons of gas. How many gallons will it need to travel 84 miles?

3.   A piece of equipment that weighs 270 pounds on Earth would weigh 45 pounds on the Theta Space Station. If an astronaut weighs 138 pounds on Earth, what would the astronaut weigh on the Theta Space Station?

4.   A van travels 168 miles on 14 gallons of gas. How many gallons will it need to travel 252 miles?

a. 21 gallons	b. 9.3 gallons
c. 20 gallons	d. 23 gallons

5.   A car travels 150 miles on 6 gallons of gas. How many gallons will it need to travel 275 miles?

        a. 11 gallons   b. 13 gallons
        c. 10 gallons   d. 3.3 gallons

6.   A piece of equipment that weighs 240 pounds on Earth would weigh 40 pounds on the Theta Space Station. If an astronaut weighs 150 pounds on Earth, what would the astronaut weigh on the Theta Space Station?

7.   A van travels 180 miles on 12 gallons of gas. How many gallons will it need to travel 435 miles?

8.   A bus travels 200 miles on 8 gallons of gas. How many gallons will it need to travel 650 miles?

9.   A van travels 180 miles on 12 gallons of gas. How many gallons will it need to travel 285 miles?

        a. 17 gallons   b. 19 gallons
        c. 18 gallons   d. 7.6 gallons

10.   A car travels 200 miles on 10 gallons of gas. How many gallons will it need to travel 300 miles?

        a. 6.7 gallons   b. 15 gallons
        c. 13 gallons   d. 14 gallons

11.   A van travels 96 miles on 8 gallons of gas. How many gallons will it need to travel 156 miles?

12.   A piece of equipment that weighs 300 pounds on Earth would weigh 50 pounds on the Theta Space Station. If an astronaut weighs 150 pounds on Earth, what would the astronaut weigh on the Theta Space Station?

13.   A van travels 60 miles on 4 gallons of gas. How many gallons will it need to travel 90 miles?

        a. 4 gallons   b. 5 gallons
        c. 6 gallons   d. 2.7 gallons

14.   A car travels 72 miles on 6 gallons of gas. How many gallons will it need to travel 360 miles?

        a. 32 gallons   b. 30 gallons
        c. 29 gallons   d. 1.2 gallons

15. A bus travels 300 miles on 12 gallons of gas. How many gallons will it need to travel 425 miles?

      a. 15 gallons   b. 16 gallons
      c. 17 gallons   d. 8.5 gallons

16. A piece of equipment that weighs 240 pounds on Earth would weigh 40 pounds on the Theta Space Station. If an astronaut weighs 174 pounds on Earth, what would the astronaut weigh on the Theta Space Station?

17. A bus travels 168 miles on 14 gallons of gas. How many gallons will it need to travel 204 miles?

18. A van travels 200 miles on 10 gallons of gas. How many gallons will it need to travel 480 miles?

      a. 22 gallons   b. 24 gallons
      c. 4.2 gallons   d. 23 gallons

19. A car travels 150 miles on 6 gallons of gas. How many gallons will it need to travel 525 miles?

      a. 21 gallons   b. 1.7 gallons
      c. 20 gallons   d. 23 gallons

20. A motorcycle travels 350 miles on 5 gallons of gasoline. What is the average miles per gallon?

21. Phoenix is 90 miles from Tucson. What is the speed of your bicycle if it takes 6 hours?

22. A van averages 8 miles to a gallon of gasoline. When pulling a car, it only gets 6 miles to a gallon. What is the difference in distance if the van only holds 40 gallons of gasoline?

23. A truck travels 200 miles on 20 gallons of diesel fuel. How many gallons will it need to travel 300 miles?

24. Spin can travel 400 miles on a tank of gasoline. He gets 20 miles per gallon regularly. With that mpg, how many gallons of gasoline will he need to travel 640 miles?

      a. 30 gallons   b. 31 gallons
      c. 32 gallons   d. 33 gallons

LESSON 42 – Distance-Rate-Time

Distance = Rate × Time

Example 1: Sally travels 300 miles in 6 hours. Her average speed, in miles per hour (mph), is what?

Distance = Rate × Time
300 miles = Rate × 6 hours
$$\frac{300 \text{ miles}}{6 \text{ hours}} = \frac{\text{Rate} \times 6 \text{ hours}}{6 \text{ hours}}$$
$$\frac{50 \text{ miles}}{1 \text{ hour}} = \text{Rate}$$
Solution: 50 mph

Example 2: It is 240 miles from Batavia to Jamestown. If it took Lenny 8 hours to make the **round trip**, what was his speed in mph?

(240 miles one way)(2) = round trip

(240 miles) • 2 = 8 hours • Rate
480 = 8R
60 = R
Solution: 60 mph

Example 3: It is $4\frac{1}{2}$ miles to the end of the trail. Kendra bikes to and back in 60 minutes. What is her speed per hour?

$(4\frac{1}{2})(2) = 1R$ 60 minutes = 1 hour
$9 = R$
Solution: 9 mph

Example 4: Joel travels 45 mph on his trip to work and back. It takes him 40 minutes to get to work. How far does he travel in his car each day?

40 minutes = $\frac{40}{60}$ hours

$$D = \frac{45 \text{ miles}}{1 \text{ hour}} \cdot \frac{40 \text{ hours}}{60} \cdot 2 = \frac{45 \cdot 40 \cdot 2}{60}$$
D = 60 miles
Solution: 60 miles

Practice problems on distance-rate-time.

1. It was 162 miles from Persy to Rimrock. If the trip took 6 hours, what was Mary's average speed in miles per hour (mph)?

2. It was 84 miles from Benton to Persy. If the **round trip** took 7 hours, what was Jane's average speed in mph?

3. It is $1\frac{4}{5}$ miles to the end of the trail. If Keenan walks to the end and back in 60 minutes, what is his average speed in mph?

4. It is $4\frac{2}{3}$ miles to the end of the trail. If Cameron bicycles to the end and back in 60 minutes, what is his average speed in mph?

5. It was 240 kilometers from Bayside to Ocean Grove. If the trip took 10 hours, what was John's average speed in kilometer per hour?

6. It was 297 miles from Medford to Newdale. If the round trip took 11 hours, what was Simone's average speed in mph?

7. It was 252 miles from Corfu to Lavonia. If the trip took 9 hours, what was Matthew's average speed in mph?

8. It was 140 miles from Newdale to Benton. If the round trip took 7 hours, what was Happy's average speed in mph?

9. Detroit to Cleveland is 110 miles. Josh takes 2 hours to drive that distance. What is his average speed in mph?

10. Courtney averages 50 mph on her 7-hour drive to Florida. What distance did she cover?

11. Aaron decides to travel on I-95 to Camden, South Carolina. If he travels an average speed of 60 mph, how long will it take him to travel the 450 miles?

12. It takes $6\frac{1}{2}$ hours to travel 325 miles. What is the average speed in mph?

13. Luke traveled 180 miles in 3 hours. What was his average speed in mph?

14. On I-95, Mrs. Jones travels at 65 miles per hour (mph). At that speed, how many miles does she travel in 8 hours?

15. From Raleigh to Orlando, it is 744 miles. Josh averages 62 mph. How long will it take him to cover that distance?

LESSON 43 – Fractional Part Word Problems

Example 1: May ate $\frac{1}{7}$ of the 21 cookies. How many cookies did she eat?

$$\frac{1}{7} \times 21 = 3 \text{ cookies}$$

Example 2: A group rented a bus to go fishing. Only $\frac{1}{3}$ of the group went and only $\frac{1}{5}$ of those paid in advance. What part of the total group paid in advance?

$$\frac{1}{3} \cdot \frac{1}{5} = \frac{1}{15}$$

Example 3: It is $3\frac{3}{4}$ miles to the end of the trail. Bruce bikes to the end and back in 60 minutes. What is his average speed in miles per hour (mph)?

D = RT
60 minutes = 1 hour

$$(3\frac{3}{4})(2) = 1R$$
$$\frac{15}{_{2}4} \cdot \frac{^{1}2}{1} = R$$
$$\frac{15}{2} = R$$

Solution: $7\frac{1}{2}$ mph

Practice problems on fractional part word problems.

1. It is $5\frac{1}{2}$ miles to the end of the trail. If Albert bicycles to the end and back in 60 minutes, what is his average speed in mph?

2. It is $2\frac{1}{3}$ miles to the end of the trail. If Keenan runs to the end and back in 60 minutes, what is his average speed in mph?

3. It is $1\frac{1}{3}$ miles to the end of the trail. If Marie walks to the end and back in 60 minutes, what is his average speed in mph?

4. Martha ate $\frac{1}{5}$ of the 25 cookies. How many cookies did she eat?

5. A ski group rented a bus to the ski slopes. Only $\frac{2}{9}$ of the group went on the trip and only $\frac{1}{6}$ of those paid in advance. What part of the total group paid in advance?

    a. $\frac{1}{5}$ b. $\frac{3}{4}$ c. $\frac{1}{27}$ d. 3

6. A ski group rented a bus to the ski slopes. Only $\frac{7}{10}$ of the group went on the trip and only $\frac{5}{9}$ of those paid in advance. What part of the total group paid in advance?

    a. $\frac{14}{9}$ b. $\frac{12}{19}$ c. $\frac{50}{63}$ d. $\frac{7}{18}$

7. Christa ate $\frac{1}{10}$ of the 70 cookies. How many cookies did she eat?

8. Florry missed $\frac{1}{5}$ of the 30 questions. How many questions did she miss?

9. A ski group rented a bus to the ski slopes. Only $\frac{7}{10}$ of the group went on the trip and only $\frac{3}{7}$ of those paid in advance. What part of the total group paid in advance?

    a. $\frac{10}{17}$ b. $\frac{10}{3}$ c. $\frac{30}{49}$ d. $\frac{3}{10}$

10. It is $3\frac{1}{5}$ miles to the end of the trail. If Adele runs to the end and back in 60 minutes, what is his average speed in mph?

11. It is $2\frac{2}{3}$ miles to the end of the trail. If Adele runs to the end and back in 60 minutes, what is his average speed in mph?

12. It is $5\frac{1}{3}$ miles to the end of the trail. If Keenan bicycles to the end and back in 60 minutes, what is his average speed in mph?

13. Charlie ate $\frac{1}{3}$ of the cookies in a jar. He ate 12 cookies. How many cookies were originally in the jar?

14. One-half of a group paid cash for their ticket. Of that group which paid cash, $\frac{3}{5}$ of them were men. There were 50 in the group. How many men paid cash for their ticket?

15. Flossy likes to bike to work and back home. It takes her 45 minutes to bike each way and the distance is 6 miles. What is her average speed in mph?

LESSON 44 – More Ratio Problems

Example 1:   If 3 pounds of grains cost $0.78, how much would 36 pounds of grains cost at the same rate?

$$\frac{3 \text{ pounds}}{\$0.78} = \frac{36 \text{ pounds}}{\$x} \quad \text{cross multiply}$$

$$(3 \text{ pounds}) \times (\$x) = (36 \text{ pounds}) \times (\$0.78)$$

$$\frac{\cancel{3 \text{ pounds}} \times (\$x)}{\cancel{3 \text{ pounds}}} = \frac{\overset{12}{\cancel{36 \text{ pounds}}} \times (\$0.78)}{\underset{1}{\cancel{3 \text{ pounds}}}}$$

$x = 12 \times (\$0.78) = \$9.36$

Solution: 36 pounds of grains cost $9.36

Example 2:   In a backyard, the ratio of birds to squirrels is 7 to 2. If there are 14 squirrels in the backyard, how many birds are there?

$$\frac{7 \text{ birds}}{2 \text{ squirrels}} = \frac{N \text{ birds}}{14 \text{ squirrels}}$$

$$(2 \text{ squirrels})(N \text{ birds}) = (7 \text{ birds})(14 \text{ squirrels})$$

$$\frac{\cancel{2 \text{ squirrels}} \cdot N \text{ birds}}{\cancel{2 \text{ squirrels}}} = \frac{7 \text{ birds} \cdot \overset{7}{\cancel{14 \text{ squirrels}}}}{\underset{1}{\cancel{2 \text{ squirrels}}}}$$

N birds = 7 birds $\cdot$ 7 = 49 birds

Solution: 49 birds

Example 3:   The ratio of good scores to bad scores are 3 to 1. If there are 18 good scores, how many were bad?

$$\frac{3 \text{ good scores}}{1 \text{ bad score}} = \frac{18 \text{ good scores}}{x \text{ bad scores}}$$

$3x = 18$

$x = 6$

Solution: 6 bad scores

Practice on ratio problems.

1. If 15 kilograms (kg) of seed cost $42, how much would 25 kg cost at the same rate?

2. A lump of taffy was dropped into the stretching machine. Twenty seconds later it was a rope of taffy 3.5 feet (ft) long. At that rate, how long would the taffy rope be in 3 minutes?

3. If 5 boxes of cherries cost $28.50, how much will 7 boxes of cherries cost?

4. Marla can exchange $260 for 400 Swiss francs. At that rate, how many dollars would a 240 franc Swiss watch cost?

5. A jumbo bean stalk grew 34 inches (in.) in 2 days. At that rate, how much would the stalk grow in 4 weeks?

    a. 39 ft 8 in.   b. 38 ft 8 in.
    c. 22 ft 8 in.   d. 4 ft 8 in.

6. The ratio of serpents to doves was 16 to 19. If there were 208 serpents, how many doves were there?

7. The cookie recipe called for oatmeal and raisins in the ratio of 3 to 1. If 5 cups of oatmeal were called for, how many cups of raisins were needed?

8. The ratio of employees to computers in a certain company is 10 to 4. If there are 70 employees in the company, how many computers are there?

    a. 27   b. 280   c. 28   d. 23

9. The ratio of good guys to bad guys was 5 to 1. If there were 55 good guys, how many bad guys were there?

10. It was 140 km from Rimrock to Medford. Matthew raced to Medford and idled back to Rimrock. If the round trip took 7 hours, what was his average speed in km/hr?

11. It is $2\frac{2}{5}$ miles to the end of the trail. If Adele runs to the end and back in 60 minutes, what is her average speed in mph?

12. Brand X cost $4.06 for 14 ounces. Brand Y cost 1¢ more per ounce. What is the cost of 8 ounces of Brand Y?

13. A single-serving carton of yogurt cost 67¢. A case of 8 single-serving cartons cost $4.80. How much is saved per carton by buying the yogurt by the case?

14. Brand X cost $18.48 for 84 ounces. Brand Y cost 5¢ more per ounce. What is the cost of 13 ounces of Brand Y?

15. The ratio of villains to heroes was 7 to 2. If there were 147 villains, how many heroes were there?
    a. 2   b. 14   c. 145   d. 42

16. The ratio of good guys to bad guys was 3 to 2. If there were 26 bad guys, how many good guys were there?
    a. 39   b. 3   c. 6   d. 23

17. The ratio of fairies to dwarfs was 6 to 19. If there were 144 fairies, how many dwarfs were there?

18. The ratio of sailors to pirates was 8 to 9. If there were 234 pirates, how many sailors were there?

19. The ratio of cops to robbers was 3 to 13. If there were 54 cops, how many robbers were there?
    a. 39   b. 234   c. 13   d. 41

20. The ratio of villains to heroes was 4 to 5. If there were 70 heroes, how many villains were there?
    a. 66   b. 20   c. 4   d. 56

21. The ratio of fairies to dwarfs was 5 to 4. If there were 95 fairies, how many dwarfs were there?

22. The ratio of rascals to good guys was 3 to 2. If there were 34 good guys, how many rascals were there?

23. The ratio of sailors to pirates was 13 to 14. If there were 260 sailors, how many pirates were there?

24. The ratio of good guys to bad guys was 1 to 4. If there were 44 bad guys, how many good guys were there?

25. The ratio of cops to robbers was 3 to 13. If there were 51 cops, how many robbers were there?

    a. 45   b. 94   c. 221   d. 5

26. The ratio of flowers to weeds was 14 to 3. If there were 60 weeds, how many flowers were there?

    a. 42   b. 46   c. 14   d. 280

27. The ratio of cops to robbers was 5 to 4. If there were 120 cops, how many robbers were there?

    a. 4   b. 116   c. 20   d. 96

28. The ratio of good guys to bad guys was 5 to 9. If there were 75 good guys, how many bad guys were there?

    a. 66   b. 135   c. 9   d. 45

29. The ratio of rascals to good guys was 2 to 13. If there were 260 good guys, how many rascals were there?

30. The ratio of fairies to dwarfs was 15 to 8. If there were 270 fairies, how many dwarfs were there?

31. The ratio of cops to robbers was 13 to 12. If there were 252 robbers, how many cops were there?

    a. 273   b. 239   c. 156   d. 13

32. The ratio of flowers to weeds was 19 to 16. If there were 323 flowers, how many weeds were there?

    a. 307   b. 304   c. 16   d. 272

33. The ratio of fairies to dwarfs was 7 to 6. If there were 102 dwarfs, how many fairies were there?

34. The ratio of good guys to bad guys was 17 to 2. If there were 204 good guys, how many bad guys were there?

LESSON 45 – Percent Story Problems

Remember: Cannot multiply or divide by a percent. Must change the percent to a decimal or fraction.

Example 1:   Felix bought a new I-Pod for $126.00. The sales tax is 4%. How much was the tax?

       Step 1:   Convert 4% to a decimal:

$$\frac{4}{100} = 0.04$$

       Step 2:   Multiply: $126 × 0.04 = $5.04

       Solution: $5.04 tax

Example 2:   Last month the phone bill was $75. This month the phone bill is $90. What percent of last month's phone bill is this month's phone bill?

       Step 1:   Divide this month's bill by last month's bill: $\frac{90}{75} = 1.2$

       Step 2:   Convert 1.2 to a percent:
                       1.2 × 100 = 120%

       Solution: 120%

Example 3:   A local newspaper's circulation decreased from 6250 in 2003 to 5675 in 2004. What percent decrease is that?

       Step 1:   Find the difference in circulation: 6250 – 5675 = 575

       Step 2:   Divide the difference by 2003's circulation: $\frac{575}{6250} = 0.092$

       Step 3:   0.092 × 100 = 9.2%
       Solution: 9.2% decrease

Practice on percent story problems.

1. The sales tax rate is 3%. Jasmine bought a pair of hiking boots advertised for $49. How much sales tax did she pay?

2. The sales tax rate is 6%. Mai Li bought a pair of rock climbing shoes advertised for $87. How much sales tax did she pay?

3. Last month's phone bill was $75. This month's phone bill is $96. What percent of last month's phone bill is this month's phone bill?

4. Last month's phone bill was $75. This month's phone bill is $132. What percent of last month's phone bill is this month's phone bill?

5. Last month's phone bill was $80. This month's phone bill is $104. What percent of last month's phone bill is this month's phone bill?

6. In 2000 the circulation of a local newspaper was 9800. In 2001 its circulation was 6860. What percent decrease is this?
   a. 23%   b. 25%   c. 22%   d. 30%

7. The sales tax rate is 7%. Alex bought a pair of track shoes advertised for $43. How much sales tax did he pay?

8. In 2000 the circulation of a local newspaper was 29,400. In 2001 its circulation was 19,110. What percent decrease is this?
   a. 30%   b. 35%   c. 31%   d. 33%

9. The sales tax rate is 5%. Jasmine bought a pair of track shoes advertised for $67. How much sales tax did she pay?

10. Last month's phone bill was $150. This month's phone bill is $201. What percent of last month's phone bill is this month's phone bill?

11. The sales tax in New York City is 8.5%. If you bought a jacket for $50, how much did the jacket cost with the tax?

12. What is the percentage increase in a heating bill from $200 in December to $325 in January?

13. Advertising in the local newspaper decreased 40% in January from December. If December's advertising was $20,000, what revenue did January bring to the newspaper owner?

LESSON 46 – More Difficult Ratio Problems

Example 1: The ratio of minnows to guppies is 7 to 5. There are 1236 minnows and guppies in the aquarium.

    a.   How many minnows are in the aquarium.

$$\frac{\text{number of minnows}}{\text{total minnows and guppies}} = \frac{\text{number of minnows}}{\text{total minnows and guppies}}$$

$$\frac{7}{7 + 5} = \frac{M}{1236}$$

$$\frac{7}{12} = \frac{M}{1236}$$

$12 \cdot M = 7 \cdot 1236$

$12M = 8652$

$M = 721$

Solution: 721 minnows

    b.   How many guppies are there?

$1236 - 721 = 515$

Solution: 515 guppies

Example 2: The ratio of advanced swimmers to beginners in the YMCA pool is 3 to 11. If there are 70 people in the pool, how many are beginners?

$$\frac{\text{beginners}}{\text{total beginners and advanced}} = \frac{\text{beginners}}{\text{total beginners and advanced}}$$

$$\frac{11}{3 + 11} = \frac{B}{70}$$

$$\frac{11}{14} = \frac{B}{70}$$

$14 \cdot B = 11 \cdot 70$

$14B = 770$

$B = 55$

Solution: 55 beginners

Practice problems on more difficult ratio problems.

1. The ratio of frogs to toads was 5 to 4. If there were 1008 frogs and toads in all, how many were frogs?

2. Farmer Ivanov wanted to plant his orchards with peaches and plums in the ratio of 5 acres to 4 acres. If he had 612 acres, how many should be planted with plums?

3. Farmer Nguyen wanted to plant his orchards with almonds and walnuts in the ratio of 6 hectares to 11 hectares. If he has 1683 hectares, how many should be planted with almonds?

4. The ratio of advanced skiers to novice skiers on the ski slope was 11 to 2. If there were 715 skiers on the ski slope, how many were advanced skiers?
    a. 605 advanced skiers   b. 55 advanced skiers
    c. 110 advanced skiers   d. 13 advanced skiers

5. The ratio of advanced skiers to novice skiers on the ski slope was 5 to 13. If there were 900 skiers on the ski slope, how many were novice skiers?
    a. 250 novice skiers   b. 18 novice skiers
    c. 650 novice skiers   d. 50 novice skiers

6. Farmer Mendoza wanted to plant his orchards with almonds and walnuts in the ratio of 13 acres to 3 acres. If he has 1040 acres, how many should be planted with walnuts?

7. The ratio of frogs to toads was 17 to 6. If there were 4922 frogs and toads in all, how many were toads?

8. The ratio of frogs to toads was 12 to 7. If there were 4446 frogs and toads in all, how many were frogs?

9. Farmer Nguyen wanted to plant her orchards with peaches and plums in the ratio of 5 acres to 8 acres. If she had 884 acres, how many should be planted with plums?

10. The ratio of frogs to toads was 16 to 9. If there were 5725 frogs and toads in all, how many were frogs?

11.   The ratio of frogs to toads was 8 to 5. If there were 2626 frogs and toads in all, how many were toads?

12.   The ratio of frogs to toads was 5 to 7. If there were 1800 frogs and toads in all, how many were frogs?

13.   The ratio of advanced skiers to novice skiers on the ski slope was 16 to 9. If there were 1475 skiers on the ski slope, how many were advanced skiers?
            a. 59 advanced skiers   b. 25 advanced skiers
            c. 531 advanced skiers   d. 944 advanced skiers

14.   The ratio of advanced skiers to novice skiers on the ski slope was 4 to 3. If there were 343 skiers on the ski slope, how many were novice skiers?
            a. 147 novice skiers   b. 49 novice skiers
            c. 7 novice skiers   d. 196 novice skiers

15.   The ratio of advanced skiers to novice skiers on the ski slope was 17 to 14. If there were 1767 skiers on the ski slope, how many were advanced skiers?
            a. 969 advanced skiers   b. 31 advanced skiers
            c. 798 advanced skiers   d. 57 advanced skiers

16.   The ratio of advanced skiers to novice skiers on the ski slope was 3 to 2. If there were 250 skiers on the ski slope, how many were advanced skiers?
            a. 100 advanced skiers   b. 150 advanced skiers
            c. 5 advanced skiers   d. 50 advanced skiers

17.   The ratio of frogs to toads was 6 to 13. If there were 4788 frogs and toads in all, how many were toads?

18.   The ratio of frogs to toads was 4 to 19. If there were 2645 frogs and toads in all, how many were frogs?

19.   Farmer Mendoza wanted to plant his orchards with apricots and almonds in the ratio of 7 acres to 11 acres. If he has 1080 acres, how many should be planted with apricots?

20.   The ratio of advanced skiers to novice skiers on the ski slope was 17 to 8. If there were 900 skiers on the ski slope, how many were novice skiers?
            a. 288 novice skiers   b. 612 novice skiers
            c. 36 novice skiers   d. 25 novice skiers

21.   The ratio of advanced skiers to novice skiers on the ski slope was 11 to 14. If there were 875 skiers on the ski slope, how many were advanced skiers?
            a. 25 advanced skiers   b. 35 advanced skiers
            c. 490 advanced skiers   d. 385 advanced skiers

LESSON 47 – More on Decimals, Percents, and Fractions

Example 1: Find the missing two answers as a decimal, fraction, or percent.

	Percent	Decimal	Fraction
a.		0.0406	
b.			$\dfrac{203}{200}$
c.	93.75%		
d.		20.125	
e.			$\dfrac{1}{16}$
f.	$32\frac{2}{5}\%$		
g.		3.16	
h.			$\dfrac{139}{800}$
i.	$53\frac{3}{4}\%$		
j.		2.054	

Solutions:

	Percent	Decimal	Fraction
a.	$4\frac{3}{50}\%$	0.0406	$\dfrac{203}{5000}$
b.	**101.5%**	**1.015**	$\dfrac{203}{200}$
c.	93.75%	**0.9375**	$\dfrac{15}{16}$
d.	**2012.5%**	20.125	$\dfrac{161}{8}$
e.	$6\frac{1}{4}\%$	**0.0625**	$\dfrac{1}{16}$
f.	$32\frac{2}{5}\%$	**0.324**	$\dfrac{81}{250}$
g.	**316%**	3.16	$\dfrac{79}{25}$
h.	$17\frac{3}{8}\%$	**0.17375**	$\dfrac{139}{800}$
i.	$53\frac{3}{4}\%$	**0.5375**	$\dfrac{43}{80}$
j.	$205\frac{2}{5}\%$	2.054	$\dfrac{1027}{500}$

Example 2:   Write each percent as a decimal.

a. $9\frac{1}{4}\%$ b.   0.03%   c. $\frac{2}{5}\%$ d.   5.2%

Solutions:

a.   0.0925   b.   0.0003   c.   0.004   d.   0.052

Practice problems on more decimals, fractions, and percents.

1.   Write each percent as a decimal number.
     a. 3.1%   b. 0.005%   c. $10\frac{1}{4}\%$ d. $\frac{9}{10}\%$

2.   Write each percent as a decimal number.
     a. 4.9%   b. 0.017%   c. $31\frac{9}{20}\%$ d. $\frac{1}{2}\%$

3.   Write each percent as a decimal number.
     a. 7.2%   b. 0.064%   c. $93\frac{1}{5}\%$ d. $\frac{1}{10}\%$

4.   Write each percent as a decimal number.
     a. 2.1%   b. 0.002%   c. $16\frac{11}{20}\%$ d. $\frac{13}{20}\%$

5.   Convert each percent to a decimal number.
     a. 3.41%   b. $99\frac{1}{2}\%$ c. $\frac{3}{5}\%$

6.   Convert each percent to a decimal number.
     a. 5.08%   b. $3\frac{3}{4}\%$ c. $\frac{1}{20}\%$

7.   Convert each percent to a decimal number.
     a. 8.29%   b. $81\frac{1}{4}\%$ c. $\frac{9}{20}\%$

8.   Write each decimal number as a percent.
     a. 0.032   b. 0.0001   c. 0.1455   d. 0.005

9.   Write each decimal number as a percent.
     a. 0.071   b. 0.00032 c. 0.4915   d. 0.003

10. Express each fraction or mixed number as a percent.
     a. $\frac{4}{5}$ b. $3\frac{11}{25}$ c. $\frac{51}{50}$ d. $\frac{9}{250}$

11. Express each of the following as a percent.
     a. 9   b. $\frac{13}{20}$ c. $\frac{1}{25}$ d. $2\frac{1}{5}$ e. $\frac{5}{12}$ f. $\frac{31}{50}$

12. Express each fraction or mixed number as a decimal.
     a. $\frac{3}{20}$ b. $\frac{9}{50}$ c. $\frac{16}{25}$ d. $1\frac{3}{4}$

13. Express each decimal as a percent.
     a. 0.49   b. 1.23   c. 0.076   d. 4.56   e. 0.987

14. Express each decimal as a percent.
     a. 0.87   b. 0.04   c. 1.37   d. 0.064

15. Convert each percent to a fraction.
     a. 3.41%   b. $99\frac{1}{2}$%   c. $\frac{3}{5}$%

16. Convert each percent to a fraction.
     a. 5.08%   b. $3\frac{3}{4}$%   c. $\frac{1}{20}$%

17. convert each percent to a fraction.
     a. 8.29%   b. $81\frac{1}{4}$%   c. $\frac{9}{20}$%

18. Express each fraction as a percent.
     a. $\frac{3}{20}$ b. $\frac{9}{50}$ c. $\frac{16}{25}$

19. Express each mixed number as a percent.
     a. $1\frac{3}{4}$ b. $3\frac{1}{2}$ c. $15\frac{1}{4}$

20. Express each mixed number as a percent.
     a. $2\frac{3}{8}$ b. $5\frac{1}{8}$ c. $8\frac{7}{20}$

LESSON 48 – Algebraic Addition

When adding signed numbers, if the signs are all the same, keep that sign and add them. If the signs are different, take the sign of largest number, and then subtract.

Example 1: $(+6) + (+3) = +9$

Example 2: $(-7) + (-3) = -10$

Example 3: $(-6) + (+4) = -2$

Example 4: $(-8) + (+9) = +1$

Example 5: $(-8) + (-6) + (+9) =$
$(-14) + (+9) = -5$

Example 6: $(-2) + (+3) + (+7) + (-5) =$
$(+1) + (+7) + (-5) =$
$(+8) + (-5) = +3$
or rearrange the addends
$(-2) + (+3) + (+7) + (-5) =$
$(-2) + (-5) + (+3) + (+7) =$
$(-7) + (+10) = +3$

Example 7: $(+8) + (+9) + (+7) + (-3) + (-6) =$
$(+17) + (+7) + (-3) + (-6) =$
$(+24) + (-3) + (-6) =$
$(+21) + (-6) = +15$
Or rearrange the addends
$(+8) + (+9) + (+7) + (-3) + (-6) =$
$(+24) + (-9) = +15$

Example 8: $(-2) + (+3) + (-4) + (+5) =$
$(+1) + (-4) + (+5) =$
$(-3) + (+5) = +2$
or rearrange the addends
$(-2) + (-4) + (+3) + (+5) =$
$(-6) + (+8) = +2$

Practice on algebraic addition. Simplify the following.

1. $(-3) + (-6) + (+3) + (+8)$

2. $(-5) + (-8) + (+2) + (-9)$

3. $(-8) + (-3) + (+6) + (-7)$

4. $(-1) + (-4) + (+9) + (-2)$

5. $(-4) + (+6) + (-1) + (+4)$

6. $(-3) + (+2) + (+9)$ a. −14   b. +8   c. +14   d. −8

7. $(-2) + (+9) + (+2)$ a. −13   b. −9   c. +9   d. +13

8. $(-9) + (+7) + (+6)$ a. −22   b. +4   c. +22   d. −4

9. $(-8) + (+3) + (-4)$ a. +9   b. −9   c. −15   d. +15

10. $(-1) + (-2) + (+5) + (+8)$

11. $(-7) + (+4) + (+6) + (+7)$

12. $(-5) + (+6) + (+7) + (-6)$

13. $(-8) + (-7) + (+8) + (+4)$

14. $(-4) + (+8) + (-1) + (-2)$

15. $(-6) + (-9) + (+4) + (+1)$

16. $(-8) + (+2) + (+8)$ a. −18   b. +18   c. −2   d. +2

17. $(-3) + (+3) + (-2)$ a. −8   b. +8   c. +2   d. −2

18.   (−6) + (+6) + (−9)   a. +9   b. −9   c. +21   d. −21

19.   (−4) + (+8) + (−3)   a. −1   b. +1   c. −15   d. +15

20.   (−2) + (+9) + (+7)   a. +18   b. −18   c. +14   d. −14

21.   (−5) + (+7) + (+1)   a. +3   b. +13   c. −13   d. −3

22.   (−7) + (+5) + (+6)   a. +18   b. −4   c. +4   d. −18

23.   (−1) + (+6) + (+1) + (+4)

24.   (−8) + (+4) + (+1) + (+5)

25.   (−7) + (+2) + (−5) + (−2)

26.   (−3) + (+7) + (+3)   a. +13   b. −7   c. −13   d. +7

27.   (−9) + (+8) + (+4)   a. −3   b. −21   c. +3   d. +21

28.   6 + (−4) + (−8) + (−9)

29.   (−9) + (−8) + (−7) + 22

30.   (−10) + 12 + 13 + (−17)

31.   (−5) + (−6) + 23 + (−10)

32.   (−14) + 15 + (−16) + 25

33.   16 + (−3) + 9 + 17

LESSON 49 – Multiplying and Dividing Signed Numbers

The rules are as follows:

> When multiplying or dividing signed numbers and they have the same sign, the answer is positive. When the signs differ, the answer is negative.

Example 1: $(-2)(-4) = 8$

Example 2: $(-3)(5) = -15$

Example 3: $(5)(4) = 20$

Example 4: $(7)(-5) = -35$

Example 5: $\dfrac{20}{-4} = -5$

Example 6: $\dfrac{-32}{-8} = 4$

Example 7: $\dfrac{-64}{16} = -4$

Example 8: $\dfrac{48}{12} = 4$

Practice on multiplying and dividing signed numbers. Simplify the following.

1.   a. $-1(-4)$ b. $-9(+3)$ c. $\dfrac{-16}{-4}$ d. $\dfrac{25}{-5}$

2.   a. $-6(-7)$ b. $-4(+7)$ c. $\dfrac{-25}{-5}$ d. $\dfrac{12}{-3}$

3.   a. $14(-80)$ b. $11(+50)$ c. $\dfrac{-170}{-17}$ d. $\dfrac{450}{-9}$

4.   a. $19(-30)$ b. $14(+70)$ c. $\dfrac{-144}{-16}$ d. $\dfrac{490}{-7}$

5. $-3(-20)$ a. -60 b. 60   c. 70   d. -70

6. $-2(-22)$ a. 44   b. -44 c. 34   d. -34

7. $\dfrac{48}{-8}$ a. -5 b. -6 c. 6   d. 5

8. $\dfrac{63}{-9}$ a. -6 b. 7   c. -7 d. 6

9. $\dfrac{36}{-4}$ a. -9 b. 10   c. -10 d. 9

10. $-9(-56)$ a. -494 b. 494   c. -504 d. 504

11.   a. $16(-40)$ b. $17(+50)$ c. $\dfrac{-99}{-11}$ d. $\dfrac{150}{-5}$

12.   a. $-5(-4)$ b. $-5(+2)$ c. $\dfrac{-12}{-3}$ d. $\dfrac{15}{-3}$

13.   a. $-3(-8)$ b. $-2(+8)$ c. $\dfrac{-12}{-4}$ d. $\dfrac{8}{-2}$

14. $-5(-28)$ a. -150 b. 150   c. 140   d. -140

15. $-4(-37)$ a. 138   b. -138 c. 148   d. -148

16. $\dfrac{12}{-3}$ a. -4 b. 5   c. 4   d. -5

17. $\dfrac{20}{-4}$ a. -6 b. 6   c. 5   d. -5

18. $-8(-40)$ a. 330   b. 320   c. -330 d. -320

19.   a. $16(-80)$ b. $15(+20)$ c. $\dfrac{-104}{-13}$ d. $\dfrac{100}{-5}$

20.   a. $-8(-9)$ b. $-6(+4)$ c. $\dfrac{-20}{-5}$ d. $\dfrac{9}{-3}$

LESSON 50 – Order of Operations

Step 1:   Remove parentheses first.
Step 2:   Multiply or divide moving from left to right.
Step 3:   Add or subtract moving from left to right.

Example 1: $6 \cdot 5 + 4(3) - 8 =$
             $30 + 12 - 8 =$
             $42 - 8 = 34$

Example 2: $5(4) + 3(2) - 6 =$
             $20 + 6 - 6 =$
             $26 - 6 = 20$

Example 3: $4 - 2 + 8(5) =$
             $4 - 2 + 40 =$
             $2 + 40 = 42$

Example 4: $(5 - 3) \cdot 6 + 4(-2 + 5) =$
             $2 \cdot 6 + 4(3) =$
             $12 + 12 = 24$

Example 5: $[6 \cdot (-2)] + 5(4 - 1) + 6 + 12 =$
             $-12 + 5(3) + 6 + 12 =$
             $-12 + 15 + 6 + 12 =$
             $3 + 6 + 12 =$
             $9 + 12 = 21$

Example 6: $-(4 - 6) + 5[4 + (-7)] + 20 =$
             $-(-2) + 5(-3) + 20 =$
             $2 - 15 + 20 =$
             $-13 + 20 = 7$

Practice problems on order of operations. Simplify or evaluate the following.

1. $2 \cdot 9 - 5(4) + 7$ a. 5   b. 45   c. 88   d. 39

2. $5 + 3 \cdot 4 - 2$

3. $3 + 2 \cdot 5 - 4$

4. $2 + 3 \cdot 4 - 5$

5.	$5 \cdot 9 - 2(-8) + 7$	a. 68	b. –35	c. 36	d. –273
6.	$7 \times 2 - 6(-3) + 4$	a. 88	b. 36	c. 0	d. –28
7.	$4 \times 6 - 5(-9) + 8$	a. –4	b. –28	c. 77	d. –13
8.	$8 \cdot 5 - 7(2) + 3$	a. 57	b. –29	c. –80	d. 29
9.	$3 \cdot 7 - 4(-6) + 9$	a. 6	b. 27	c. 54	d. –45
10.	$2 + 5 \times 4 - 3$				
11.	$8 \cdot 9 - 6(5) + 4$	a. 106	b. 216	c. 124	d. 46
12.	$4 \cdot 6 - 3(-7) + 2$	a. –60	b. 47	c. 5	d. –82
13.	$2 \times 3 - 8(-9) + 5$	a. 40	b. 95	c. 83	d. –61
14.	$5 \cdot 8 - 4(-6) + 7$	a. –113	b. 20	c. 23	d. 71
15.	$-(-5) - 3(9 - 6)$	a. 14	b. –28	c. –40	d. –4
16.	$-(-6) - 9(10 - 5)$	a. –129	b. 51	c. –89	d. –39
17.	$-(-8) - 8(10 - 7)$				
18.	$6(-4 + 5)(-3 + 3) + 6$				
19.	$7(-9 + 4)(-1 + 7) + 2$				
20.	$4(-2 + 9)(-4 + 6) + 5$				
21.	$-(-9) - 7(8 - 2)$	a. 51	b. –33	c. –61	d. –49
22.	$-2(-2) - 5(11 - 7)$	a. –88	b. –60	c. 22	d. –16
23.	$-(-6) - 5(10 - 8)$				
24.	$-(-7) - 6(6 - 4)$	a. –5	b. –33	c. –53	d. 19
25.	$-(-9) - 4(9 - 6)$	a. –3	b. –33	c. –51	d. 21
26.	$-(-8) - 3(8 - 6)$				

27. $-2(4 + 2)(9 + 4) + 8$

28. $-8(-2 + 4) - 3(4 - 3)$

29. $-4(-3 + 6)$ a. -6 b. -12 c. 18 d. 12

30. $44 - 2(5 + 3)$

31. $40 - 3(7 - 4)$

32. $46 - 4(6 + 2)$

33. $-4(-5 + 4)$ a. 24 b. -4 c. -16 d. 4

34. $2(-4 + 2)$ a. 10 b. -4 c. -6 d. 4

35. $-2(-7 + 5) - 3(2 - 8)$

36. $-6(4 + 3) - 7(-7 - 1)$

37. $-2(-6 + 7)(7 + 6) + 8$

38. $5(9 + 1)(4 + 7) + 2$

39. $8(-2 + 9) - 2(-6 - 9)$

40. $3(-4 + 3)$ a. 3 b. -3 c. 15 d. -9

41. $3(2 + 6) - 2(-7 - 2)$

42. $7(9 + 7) - 3(-3 - 9)$

43. $-2(-5 + 2)$ a. 12 b. 6 c. -8 d. -6

44. $3(-5 + 4) - 9(-7 - 6)$

45. $-2(3 - 4) + 3(2 + 5)$

46. $-5(3 - 1) + 4(1 - 3)$

47. $4(7 - 2)$

48. $2(6 - 8) - 5(7 - 10)$

LESSON 51 – Evaluating a Variable Expression

To solve these expressions, replace the variable with the given constant and follow the rules for order of operations.

Example 1:   a + 3b   when a = 2 and b = 4

$$a + 3b =$$
$$2 + 3(4) =$$
$$2 + 12 = 14$$

Example 2:   2x + 4y   when x = –1 and y = 5

$$2x + 4y =$$
$$2 \cdot (-1) + 4 \cdot 5 =$$
$$-2 + 20 = 18$$

Example 3:   mn – 4n   when m = 5 and n = 1

$$mn - 4n =$$
$$5(1) - 4 \cdot 1 =$$
$$5 - 4 = 1$$

Example 4:   2ab – 3c   when a = 4, b = –3, and c = 2

$$2ab - 3c =$$
$$2(4)(-3) - 3(2) =$$
$$8(-3) - 3(2) =$$
$$-24 - 6 = -30$$

Example 5:   efg + 2fg   when e = 4, f = 3, and g = 2

$$efg + 2fg =$$
$$(4)(3)(2) + 2(3)(2) =$$
$$12(2) + 6(2) =$$
$$24 + 12 = 36$$

Example 6:   5(x + y)(x – y) + xy   when x = 3 and y = 2

$$5(x + y)(x - y) + xy =$$
$$5(3 + 2)(3 - 2) + 3(2) =$$
$$5(5)(1) + 6 =$$
$$25(1) + 6 =$$
$$25 + 6 = 31$$

Practice problems on evaluating variable expressions. Evaluate the following.

1. $2e - f$ if e = 15 and f = 1
    a. 13   b. 31   c. 28   d. 29

2. $5u + v$ if u = 3 and v = 14
    a. 29   b. 85   c. 73   d. 1

3. $-(u + t)(u - t)$ if t = 3 and u = –5

4. $-(j - k)(j + k)$ if j = –6 and k = 4

5. $-h(-f + g) - fg$ if f = –5, g = –5, and h = –4

6. $-f(-d + e) - de$ if d = 4, e = –1, and f = 3

7. $-h(f + g)$ if f = 1, g = –4, and h = –3

8. $mn + 2m$ if m = –6 and n = 3

9. $bc - 3c$ if b = –3 and c = 7

10. How many of these products are positive?

    i. $(+3)(-4)(-8)(-6)$
    ii. $(-1)(-7)(+8)(+3)(-1)(-1)$
    iii. $(+5)(+1)(-6)$

        a. 2   b. 3   c. 1   d. 0

11. How many of these products are negative?

    i. $(-2)(-3)(+3)(+7)$
    ii. $(-6)(-1)(+7)(-1)(-7)(+7)$
    iii. $(+1)(+5)(+8)$

        a. 3   b. 2   c. 1   d. 0

12. $de - 5e$ if d = 5 and e = –6

13. $-v(t - u)$ if t = 2, u = –5, and v = –2

14. $-c(-a + b) - ab$ if a = 1, b = 4, and c = 3

116

15. $-(j - k)(j + k)$ if $j = 2$ and $k = 7$

16. $3e - f$ if $e = 5$ and $f = 7$
    a. 6   b. 16   c. 8   d. 22

17. $5u + v$ if $u = 4$ and $v = 3$
    a. 23   b. 35   c. 19   d. 17

18. $4t + u$ if $t = 8$ and $u = 11$
    a. 21   b. 43   c. 52   d. 76

19. $-(c - d)(c + d)$ if $c = 5$ and $d = 7$

20. $-g(-e + f) - ef$ if $e = 1$, $f = -2$, and $g = -1$

21. $-w(-u + v) - uv$ if $u = -2$, $v = 1$, and $w = -7$

22. $-r(p - q)$ if $p = 1$, $q = -2$, and $r = 5$

23. $-c(a + b)$ if $a = 5$, $b = 6$, and $c = 1$

24. $qr + 2q$ if $q = -7$ and $r = -3$

25. $-(e - f)(e + f)$ if $e = 6$ and $f = -1$

26. $xy + (-z)$ if $x = 4$, $y = 5$, and $z = -4$

27. $a(-b) - ac$ if $a = 3$, $b = -2$, and $c = 12$

28. $(3d + 4e) - 2f$ if $d = -2$, $e = -4$, and $f = -6$

29. $(4x - 5y)(z - x)$ if $x = 5$, $y = -2$, and $z = 8$

30. $-a(b + c) - c(b + a)$ if $a = 4$, $b = -3$, and $c = 7$

31. $pq(rs)$ if $p = 2$, $q = 3$, $r = 5$, and $s = -3$

32. $3xy(-4z)$ if $x = 3$, $y = -2$, and $z = -5$

33. $(ab) - 2(c)d$ if $a = 4$, $b = 3$, $c = 1$, and $d = 6$

34. $2(e - f) + (gh)$ if $e = 8$, $f = 5$, $g = 2$, and $h = 4$

35. $(5x - 2y) + z(x)$ if $x = 8$, $y = 4$, and $z = 7$

LESSON 52 – Simple Equations Using Addition and Subtraction

Example 1: $x + 5 = 8$

$$
\begin{array}{rrcr}
x & + \ 5 & = & 8 \\
  & - \ 5 & & -5 \\
\hline
x & & = & 3
\end{array}
$$

Example 2: $16 + N = 35$

$$
\begin{array}{rrcr}
16 & + \ N & = & 35 \\
-16 & & & -16 \\
\hline
 & N & = & 19
\end{array}
$$

Example 3: $a - 15 = 24$

$$
\begin{array}{rrcr}
a & - \ 15 & = & 24 \\
  & + \ 15 & & +15 \\
\hline
a & & = & 39
\end{array}
$$

Example 4: $33 - N = 78$

$$
\begin{array}{rrcr}
33 & - \ N & = & 78 \\
-33 & & & -33 \\
\hline
 & - \ N & = & 45 \\
 & N & = & -45
\end{array}
$$

Example 5: $35 - b = 14$

$$
\begin{array}{rrcr}
35 & - \ b & = & 14 \\
-35 & & & -35 \\
\hline
 & - \ b & = & -21 \\
 & b & = & 21
\end{array}
$$

Example 6: $c - 14 = 32$

$$
\begin{array}{rrcr}
c & - \ 14 & = & 32 \\
  & + \ 14 & & +14 \\
\hline
c & & = & 46
\end{array}
$$

Example 7: $9 + e = 26$

$$
\begin{array}{rcr}
9 \;+\; e &=& 26 \\
-9 && -9 \\
\hline
e &=& 17
\end{array}
$$

Example 8: $f + 13 = -56$

$$
\begin{array}{rcr}
f \;+\; 13 &=& -56 \\
-\; 13 && -13 \\
\hline
f &=& -69
\end{array}
$$

Practice problems on simple equations using addition and subtraction. Solve for the variable.

1. $N + 7 = 33$

2. $N + 6 = 40$

3. $n - 2 = 29$ a. $n = 28$ b. $n = 27$ c. $n = 32$ d. $n = 31$

4. $y - 4 = 7$ a. $y = 11$ b. $y = 4$ c. $y = 10$ d. $y = 3$

5. $j - 9 = 27$ a. $j = 36$ b. $j = 37$ c. $j = 18$ d. $j = 17$

6. $r - 9 = 15$

7. $t - 6 = 10$

8. $c - 3 = 11$

9. $N - 8 = 20$

10. $N - 8 = 18$

11. $g - 4 = 21$

12. $j - 6 = 23$ a. $j = 29$ b. $j = 17$ c. $j = 18$ d. $j = 28$

13. $N + 9 = 36$

14. $N + 8 = 39$

15. $N + 6 = 39$

16. $3 + f = 11$ a. f = 7   b. f = 16   c. f = 3   d. f = 8

17. $9 + e = 15$ a. e = 6   b. e = 9   c. e = 12   d. e = 7

18. $N + 6 = 43$

19. $c - 3 = 25$ a. c = 21   b. c = 28   c. c = 29   d. c = 22

20. $v - 7 = 9$ a. v = 17   b. v = 16   c. v = 2   d. v = 3

21. $j - 4 = 18$

22. $d - 6 = 14$

23. $N - 8 = 19$

24. $N - 9 = 28$

25. $N + 4 = -16$

26. $a - 8 = 26$

27. $17 - b = -25$

28. $6 + x = -13$

29. $4 - c = 22$

30. $16 + b = -2$

31. $D - 4 = 22$

32. $x + 5 = -15$

33. $5 - N = 20$

34. $13 - x = 6$

35. $42 - a = 50$

LESSON 53 – Simple Equations Using the Division and Multiplication Rule

Example 1: $6x = 18$

$$\frac{6x}{6} = \frac{18}{6} \quad \rightarrow \quad x = 3$$

Example 2: $4 \cdot N = 36$

$$\frac{4 \cdot N}{4} = \frac{36}{4} \quad \rightarrow \quad N = 9$$

Example 3: $5a = 8 \cdot 10$

$$5a = 80 \quad \rightarrow \quad \frac{5a}{5} = \frac{80}{5} \quad \rightarrow \quad a = 16$$

Example 4: $b \div 6 = 8$

$$b \div 6 = 8 \quad \rightarrow \quad \frac{b}{6} = 8 \quad \rightarrow \quad \frac{6 \cdot b}{6} = 8 \cdot 6 \rightarrow \quad b = 48$$

Example 5: $7 \times n = 28$

$$\frac{7 \times n}{7} = \frac{28}{7} \quad \rightarrow \quad n = 4$$

Example 6: $4e = 8 \cdot 7$

$$4e = 56 \quad \rightarrow \quad \frac{4e}{4} = \frac{56}{4} \quad \rightarrow \quad e = 14$$

Example 7:   Solve for m: $3m = 36$

$$\frac{3m}{3} = \frac{36}{3} \quad \rightarrow \quad m = 12$$

Example 8: $\frac{x}{4} = 13 \quad \rightarrow \quad \frac{4 \cdot x}{4} = 13 \cdot 4 \quad \rightarrow \quad x = 52$

Example 9: $x \div 3 = 17 \quad \rightarrow \quad \frac{x}{3} = 17 \quad \rightarrow \quad \frac{3 \cdot x}{3} = 17 \cdot 3 \quad \rightarrow \quad x = 51$

Practice problems on simple equations using the division and multiplication rules.

1. Solve for h: $7h = 56$ a. 63   b. 49   c. 8   d. 392

2. Solve for s: $6s = 12$ a. 6   b. 2   c. 18   d. 72

3. $m \times 10 = 70$

4. $12 \times c = 36$

5. Solve for p: $8p = 120$

6. Solve for s: $4s = 32$

7. $4z = 32 \cdot 5$

8. $2m = 5 \cdot 2$

9. Solve for b: $11b = 33$

10. Solve for r: $4r = 36$

11. Solve for w: $7w = 21$

12. $7f = 4 \cdot 7$

13. Solve for g: $\dfrac{g}{6} = 3$ a. 18   b. $\frac{1}{2}$ c. 2   d. $\frac{1}{18}$

14. Solve for m: $\dfrac{m}{3} = 2$ a. $\frac{2}{3}$ b. $\frac{3}{2}$ c. $\frac{1}{6}$ d. 6

15. $n \div 5 = 12$

16. $c \div 3 = 10$

17. $\dfrac{N}{5} = 7$

18. $\dfrac{N}{6} = 7$

19. $\dfrac{N}{3} = 5$

20. $j \div 8 = 10$

21. Solve for t: $\frac{t}{4} = 3$ a. $\frac{3}{4}$ b. $\frac{1}{12}$ c. 12   d. $\frac{4}{3}$

22. $x \div 9 = 11$

23. $\frac{N}{5} = 3$

24. Solve for u: $\frac{u}{3} = 7$ a. $\frac{1}{21}$ b. $\frac{3}{7}$ c. $\frac{7}{3}$ d. 21

25. $d \div 3 = 11$

26. $p \div 2 = 9$

27. Solve for f: $\frac{f}{8} = 10$ a. $\frac{4}{5}$ b. $\frac{5}{4}$ c. 80   d. $\frac{1}{80}$

28. Solve for c: $\frac{c}{2} = 6$ a. $\frac{1}{3}$ b. $\frac{1}{12}$ c. 12   d. 3

29. $4u = 10 \cdot 16$

30. $6x = 7 \cdot 18$

31. $\frac{a}{10} = 6$

32. $x \div 4 = 8$

33. $\frac{N}{5} = 32$

34. $5x = 75$

35. $3x = 6 \cdot (-3)$

36. $4x = 18$

37. $\frac{c}{7} = 8$

38. $6x = 8(4.5)$

LESSON 54 – Revisiting Order of Operations

Step 1:   Remove parenthesis and exponents from left to right.
Step 2:   Multiply or divide from left to right.
Step 3:   Add or subtract from left to right.

Example 1: $5^2 + (3 \times 7) - 4 = 25 + 21 - 4 = 46 - 4 = 42$

Example 2: $10 + 2^2 + 4 = 10 + 4 + 4 = 14 + 4 = 18$

Example 3: $10^2 - 7^2 + 5 = 100 - 49 + 5 = 51 + 5 = 56$

Example 4: $(2^3 - 3) + 5^2 \cdot 3 = (8 - 3) + 25 \cdot 3 = 5 + 25 \cdot 3 = 5 + 75 = 80$

Practice on order of operations. Simplify the following.

1. $2 + 5 \cdot 3^2$ a. 63   b. 47   c. 227   d. 289

2. $6 + 4^2$

3. $8^2 - 5 - 7 \times 4$

4. $6^2 - 2 - 2 \times 6$

5. $8^2 - 4 - 5 \cdot 2$

6. $3^3 - 2^3 - 2 \cdot 3$

7. $9 + 1^3$

8. $8 + 2^3$

9. $1^2 - 1$

10. $4^3 - 3^3 - 3 \times 4$

11. $10^2 - 4 - 2 \times 9$

12. $8 + 4 \cdot 3^2$ a. 152   b. 400   c. 44   d.   108

13. $7^2 - 5 - 8 \times 3$

14. $6^3 - 3^3 - 3 \cdot 6$

15. $5^3 - 2^3 - 2 \times 5$

LESSON 55 – Bracketology
(Removal of Brackets/Parentheses)

Example 1: $6 + (4 \cdot 3) - 5 = 6 + 12 - 5 = 18 - 5 = 13$

Example 2: $4 + (9 \cdot 2) + 7 = 4 + 18 + 7 = 22 + 7 = 29$

Example 3: $3(7 - 4) + 15 = 3(3) + 15 = 9 + 15 = 24$

Example 4: $2(5 - 2)(1 + 3) - 8 = 2(3)(4) - 8 = 6(4) - 8 = 24 - 8 = 16$

Example 5. $7(3 - 5) + 22 = 7(-2) + 22 = -14 + 22 = 8$

Example 6: $5(1 - 4)(3 - 4) + 9 = 5(-3)(-1) + 9 = (-15)(-1) + 9 = 15 + 9 = 24$

Example 7: $6 + \left(\dfrac{18}{9}\right) - 5 = 6 + 2 - 5 = 8 - 5 = 3$

Example 8: $10 + \dfrac{32}{-4} + 6 = 10 + (-8) + 6 = 2 + 6 = 8$

   or: $10 + \dfrac{32}{-4} + 6 = 10 - 8 + 6 = 2 + 6 = 8$

Example 9: $15 + \dfrac{24}{-6} - 20 = 15 + (-4) - 20 = 11 - 20 = -9$

   or: $15 + \dfrac{24}{-6} - 20 = 15 - 4 - 20 = 11 - 20 = -9$

Example 10: $\dfrac{(4 - 2)(5 - 6) + 18}{(-7) + 9} = \dfrac{2(-1) + 18}{2} = \dfrac{-2 + 18}{2} = \dfrac{16}{2} = 8$

Example 11: $\dfrac{5 + (3 - 1)(7 - 4) + 1}{4 - 7} = \dfrac{5 + (2)(3) + 1}{-3} = \dfrac{5 + 6 + 1}{-3} = \dfrac{11 + 1}{-3} = \dfrac{12}{-3} = -4$

Practice on bracketology. Simplify the following.

1. $3(-2 + 5)$ a. -1 b. 9 c. 11 d. -9

2. $2(-5 + 6)$ a. -4 b. 2 c. 16 d. -2

3. $50 - 4(6 + 2)$

4. $43 - 3(7 - 4)$

5. $4(-2 + 5)$ a. 13   b. 12   c. −12   d. −3

6. $-2(-5 + 2)$ a. 12   b. −6   c. −8   d. 6

7. $7(-6 + 5) - 7(-5 - 7)$

8. $-9(3 + 1) - 5(1 - 3)$

9. $4(7 + 6)(-5 + 3) + 5$

10. $7(6 + 4)(2 + 4) + 7$

11. $2 + \dfrac{-16}{8} + 3$

12. $\dfrac{(6 + 1) + (-5 - 5)}{-8 + (-9)}$

13. $\dfrac{(8 - 4) + (6 + 5)}{8 - (-3)}$

14. $-3 + \dfrac{6}{-2} - 7$

15. $4(-2 + 4)(8 + 2) + 8$

16. $-3(6 + 7) - 5(-8 - 4)$

17. $-4(-2 + 1)$ a. −4   b. 9   c. −7   d. 4

18. $3(-3 + 5)$ a. 6   b. 14   c. −4   d. −6

19. $40 - 2(5 - 2)$

20. $40 - 2(6 - 4)$

21. $-2(-2 + 1)$ a. 5   b. −3   c. −2   d. 2

22. $-4(9 + 4) - 7(-3 - 1)$

23. $-8(-2 + 4) - 3(4 - 3)$

LESSON 56 – Evaluate a Square Root

Example 1: $\sqrt{16} = 4$

Example 2: $\sqrt{25} = 5$

Example 3: $\sqrt{121} = 11$

Practice on evaluating a square root. Use the square root key on your calculator to find the square root. Then solve the problem using PEMDAS (order of operations).

1. $\sqrt{9}$

2. $\sqrt{36}$

3. $\sqrt{49}$

4. $\sqrt{144}$

5. $\sqrt{100}$

6. $\sqrt{81}$

7. $\sqrt{25} + \sqrt{36}$

8. $(\sqrt{4})(\sqrt{9})$

9. $\sqrt{225}$

10. $\sqrt{169}$

11. $(\sqrt{144})(\sqrt{4})$

12. $\sqrt{64}$

13. $\sqrt{121} + \sqrt{100}$

14. $\sqrt{196}$

15. $(\sqrt{4})(\sqrt{9})(\sqrt{16})$

16. $\sqrt{100} - \sqrt{36}$

17. $\sqrt{16} + \sqrt{9}$

18. $(\sqrt{25})(\sqrt{49})$

19. $2(\sqrt{4})$

20. $3(\sqrt{4}) + \sqrt{16}$

21. $\sqrt{36} - \sqrt{16}$

22. $\sqrt{49}(\sqrt{9})$

23. $\sqrt{25} + \sqrt{25} + \sqrt{25}$

24. $(\sqrt{64})(\sqrt{81})$

25. $(\sqrt{100})(\sqrt{64})$

LESSON 57 – Working with a Cube Root

Example 1: $\sqrt[3]{8} = 2$ Example 2: $\sqrt[3]{64} = 4$

Example 3: $3^3 + \sqrt[3]{125} = 27 + 5 = 32$

Practice working with a cube root. Simplify the following.

1. $\sqrt[3]{27}$

2. $\sqrt[3]{125}$

3. $4^2 + \sqrt[3]{27}$

4. $2^2 + \sqrt[3]{64}$

5. $\sqrt[3]{8} + 2^2$

6. $4^2 + \sqrt[3]{125}$

7. $3^3 + \sqrt[3]{27}$

8. $9(\sqrt[3]{8})$

9. $(\sqrt[3]{1})(\sqrt[3]{64})(\sqrt[3]{27})$

10. $(\sqrt[3]{125})(\sqrt{25})$

11. $4^3 - \sqrt[3]{64}$

12. $(18)(1^3) + \sqrt[3]{8}$

13. $6^2 + (\sqrt[3]{8})(\sqrt[3]{27})$

14. $5^2 - (\sqrt[3]{125})(\sqrt[3]{8})$

15. $(\sqrt{36} + \sqrt[3]{64}) - \sqrt{144}$

16. $\sqrt[3]{8}(\sqrt[3]{27} + \sqrt{4})$

17. $\sqrt[3]{216}$

18. $\sqrt[3]{343}$

19. $\sqrt[3]{512}$

20. $(\sqrt[3]{8})(\sqrt[3]{27})$

21. $(\sqrt[3]{125})^2$

22. $(\sqrt[3]{1})^{24}$

23. $(\sqrt[3]{27})(2 + \sqrt[3]{8})$

24. $6^2 + \sqrt[3]{216}$

25. $(\sqrt[3]{64})(\sqrt[3]{8} + \sqrt[3]{27})$

26. $5^2(x^3)$

27. $1^{10} + \sqrt[3]{1000}$

28. $(\sqrt[3]{343})(\sqrt[3]{512})$

29. $(\sqrt[3]{27})(2^2 + \sqrt{25})$

30. $(\sqrt[3]{729})(\sqrt[3]{27})$

31. $\dfrac{\sqrt{16}(\sqrt[3]{8})^2}{2^2}$

32. $\dfrac{(\sqrt[3]{64})^2}{\sqrt[3]{1000}}$

33. $(\sqrt[3]{125})(\sqrt[3]{125})$

LESSON 58 – Evaluating Exponents with Negative Bases

Example 1: $(-2)^2 = (-2)(-2) = 4$

Example 2: $(-2)^3 = (-2)(-2)(-2) = (+2)(-2) = -8$

Example 3: $(-4)^2 = (-4)(-4) = 16$

Example 4: $5^2 - (-2)^3 = 25 - (-2)(-2)(-2) = 25 - (+2)(-2) = 25 - (-8) = 25 + 8 = 33$

Example 5: $(-3)^2 + 8^2 = (-3)(-3) + (8)(8) = 9 + 64 = 73$

Example 6: $-5^2 + 15 = -(5^2) + 15 = -(5)(5) + 15 = -25 + 10 = -10$

Example 7: $-3^3 + \sqrt[3]{64} = -(3)(3)(3) + 4 = -(9)(3) = -27 + 4 = -23$

Example 8: $(\sqrt[3]{8})(-2)^3 = 2(-2)(-2)(-2) = 2(-8) = -16$

Example 9: $(\sqrt[3]{8}) + (-2)^3 = 2 + (-2)(-2)(-2) = 2 + (-8) = 2 - 8 = -6$

Example 10: $-3^4 + \sqrt[3]{125}(\sqrt[3]{64}) = -(3)(3)(3)(3) + 5(4) = -81 + 20 = -61$

Practice problems evaluating exponents with negative bases.

1. $(-3)^3$ a. 9   b. –27   c. 27   d. –9

2. $-(2)^4$ a. –16   b. 16   c. –8   d. 8

3. $-(3)^2$ a. 6   b. –9   c. 9   d. –6

4. $-2^3 - (-2)^2 - (-3)^3 + \sqrt[3]{64}$

5. $-2^4 - (-3)^2 - (-2)^4 + \sqrt[3]{8}$

6. $-3^3 - (-2)^4 - (-3)^2 + \sqrt[3]{27}$

7. $2^2 + \sqrt[3]{8}$

8. $3^3 + \sqrt[3]{27}$

9. $(-3)^3 + \sqrt[3]{27}$

10. $(-2)^3 + (-1)^4$

11. $4^2 + \sqrt[3]{64}$

12. $-3^2 - (-3)^2 - (-2)^4 + \sqrt[3]{27}$

13. $-2^2 - (-2)^3 - (-3)^2 + \sqrt[3]{8}$

14. $-2^4 - (-2)^2 - (-3)^3 + \sqrt[3]{64}$

15. $-3^2 - (-2)^2 - (-2)^4 + \sqrt[3]{27}$

16. $-2^4 - (-2)^3 - (-3)^3 + \sqrt[3]{8}$

LESSON 59 – Absolute Value

When working with absolute value problems, perform the mathematics inside the absolute value symbols (| |) first, then extract the positive answer to continue. Hopefully these examples will help.

Example 1: |3| = 3

Example 2: |–4| = 4

Example 3: 6 + |–3 + 2| = 6 + |–1| = 6 + 1 = 7

Example 4: 5 + |2 – 8| = 5 + |–6| = 5 + 6 = 11

Example 5: 3 – 8 + (6 + 4) – |10| = 3 – 8 + 10 – 10 = –5 + 10 – 10 = 5 – 10 = –5
  or 3 – 8 + (6 + 4) – |10| = 3 – 8 + 10 – 10 = –5 + 0 = –5

Example 6: 4 + |2 – 5| – |5 – 3| = 4 + |–3| – |2| = 4 + 3 – 2 = 7 – 2 = 5

Example 7: |12 – 15| = |–3| = 3

Example 8: 5 + |1 – 4| = 5 + |–3| = 5 + 3 = 8

Example 9: 10 + |2 – 8| = 10 + |–6| = 10 + 6 = 16

Example 10: |4 – 8| – | 8 – 4| = |–4| – |4| = 4 – 4 = 0

Practice on absolute value. Simplify the following.

1. |–4 + 4| – 5 – 3

2. |1 – 3| + 6 + 8

3.   –|–7 + 4| – 7 + 7

4. |1 – 5| + 5 + 7

5.   –7 + |–7 – 6| – 4   a. –4   b. 4   c. –24   d. 2

6.   2 – |–3 + 4| + 7   a. –2   b. 8   c. –8   d. 2

7.   3 + |7 + 2| – 9   a. –3   b. 7   c. 3   d. –17

8.   –(–6) – (–1) – 3

130

9. $7 - |-5 + 3| - 4$ a. 5   b. 1   c. −19   d. 19

10. $-9 + |2 - 4| + 4$ a. −3   b. −7   c. 7   d. −19

11. $4 - |3 - 3| - 8$ a. 4   b. −4   c. 18   d. −6

12. $-7 + |-4 + 3| + 5$ a. −3   b. −19   c. 19   d. −1

13. $-(-8) - |-11| + 5$

14. $-(-1) - (-5) - 3$

15. $10 - 6 - |-14| + |11 - 10 + 6|$

16. $12 - 14 - |-15| + |7 - 2 - 5|$

17. $4 - 7 - (-12) + |3 - 13 - 4|$

18. $7 - |4 + 6| - 7$ a. 16   b. −10   c. −16   d. 10

19. $|6 - 9| + 4^2$

20. $(\sqrt[3]{27}) - |-12| + 5$

21. $-|-13| + |-5^2| - (5 \cdot 2)$

22. $3^3 + |-8 - 4 - 12| - 3^2$

23. $|-3 - 5 - 9| - |3^2|$

24. $(\sqrt{49})|-2 - 3 - 8 + 1|$

25. $|4^2| - |-8| + |-16|$

26. $|-3^2 - 4 + 5 - 6|$

27. $|(-3)^2 - 4 + 5 - 6|$

28. $|(-3)^2 - 4 - 5 - 6|$

29. $|-3^2 + 4 - 5 + 6|$

30. $|(-3)^2 + 4 - 5 + 6|$

LESSON 60 – Left to Right Inside Parentheses with Lots of Signs

Remember: PEMDAS – Parentheses, Exponents, Multiply/Divide, Add/Subtract
Note: The symbols | | represents absolute value.

Example 1: $(-4) + |-\{-4\}| = -4 + |+4| = -4 + 4 = 0$

Example 2: $(6) + |-10| - (5)[-(-(-4))] = 6 + 10 - 5[-4] = 6 + 10 + 20 = 16 + 20 = 36$

Example 3: $(-[-3]) + |5(-5)| = (3) + |-25| = 3 + 25 = 28$

Example 4: $|5 - 8| + [-(-[-8])] = |-3| + [-8] = 3 + [-8] = 3 - 8 = -5$

Example 5: $(-[-(-[-6])]) + |6 - [-10]| = (+6) + |6 + 10| = 6 + |16| = 6 + 16 = 22$

Example 6: $5 - [-(7)] + (-8)(1) = 5 + 7 + (-8) = 5 + 7 - 8 = 12 - 8 = 4$

Example 7: $9 - 7 + [-(6)] - |-13| = 9 - 7 - 6 - 13 = 2 - 6 - 13 = -4 - 13 = -17$

Example 8: $|-6| - [-3(-2)] + (7)(-3) = 6 - [+6] + (-21) = 6 - 6 - 21 = 0 - 21 = -21$

Example 9: $(-8) + [-(-4)] + (4)(-2)(-1) = -8 + [+4] + (-8)(-1) = -8 + 4 + 8 =$
$-4 + 8 = 4$

Example 10: $[2] + |6 - 8| - (2)[-(-4)] = 2 + |-2| - (2)[+4] = 2 + 2 - 8 = 4 - 8 = -4$

Example 11: $(-[6]) + (5)(-3) + |-4| = (-6) + (-15) + 4 = -6 - 15 + 4 = -21 + 4 = -17$

Example 12: $[5] + (7)(-2) - [-(-[6])] = 5 + (-14) - [+6] = 5 - 14 - 6 = -9 - 6 = -15$

Practice on left to right inside parentheses with lots of signs. Simplify the following.

1. $-(-7) - (-6) + (-14)$

2. $-7 + |-6 - 8| + 3$ a. -18 b. 8 c. 10 d. -12

3. $-(-2) - (-6) + (-13)$ a. 21 b. -21 c. -5 d. -17

4. $-(-8) - (-11) - 6$ a. 13 b. 3 c. -13 d. 25

5. $-(-5) - (-7) + 15$ a. -3 b. 27 c. 3 d. 13

6.   –(–6) – (–2) + (–6)

7.   3 – 8 – |–9| + |12 – 1 + 6|

8.   4 – 2 – (–4) – [–|–2|]

9.   –[–(–2)] – {–[–(–14)]}

10.   –(–1) – |–12| + 6   a. 19   b. –5   c. –7   d. 7

11.   –[–(–2)] – {–[–(–13)]}

12.   1 – 13 – |–4| – [–(–10)]

13.   –(–6) – (–14) + 13   a. 7   b. –7   c. 5   d. 33

14.   –(–12) – (–13) – (–9)   a. –10   b. –34   c. 34   d. 16

15.   –|–(– 4)| – {–[–|–6|]}

16.   6 – 8 – (–7) – [–|–11|]

17.   4 – 7 – (–5) – [–(–2)]

18.   –[–(– 4)] – {–[–|–9|]}

19.   –|–(–11)| – {–[–|–5|]}

20.   –(–10) – (–9) + (–12)   a. –31   b. –11   c. 7   d. 31

21.   –(–11) – (–10) + 7   a. –14   b. 14   c. 8   d. 28

22.   –|–10| – |–5| + 2   a. 7   b. 17   c. 13   d. –13

23.   –[–(–3)] – {–[–|–11|]}

24.   –|–(–10)| – {–[–|–10|]}

25.   –|–(–2)| – {–[–|–6|]}

26.   10 – 1 – (–5) – [–|–9|]

27.   3 – 11 – |–14| + |2 – 13 – 10|

A Short Review of Previous Lessons in Book 1

Simplify.

1. $6(4 + 9)$

2. $(4)(6)(2 \cdot 3)$

3. $\dfrac{4(9 + 4) + 8}{3(6 - 4)}$

4. $-(-8) - 3(8 - 7)$

5. $-(4 - 6) + 5[4 + (-7)]$

6. $\dfrac{(-6)(4)}{(2)(6) - (6 + 4)}$

7. $7(2 \cdot 3 + 6)$

8. $4(2 + 4 - 7)$

9. $-(2) \div (8 - 6)$

10. $[6 \cdot (2)] + 5(4 - 1)$

11. $4 - 2 + 4(5)$

12. $7(3 - 5) + 20$

13. $8 + \dfrac{18}{2} - 6$

14. $\dfrac{(3 - 1)(5 - 7) + 20}{4 + (-6)}$

15. $-8(-4 + 2) + 5$

16. $\sqrt[3]{64} + \sqrt[3]{8}$

17. $(\sqrt{81})(\sqrt{64})$

18. $-(\sqrt{36}) + |-10|$

19. $(\sqrt[3]{27})(\sqrt{25})$

20. $|-17| + |32|$

21. $\sqrt{8^2} - \sqrt[3]{125}$

22. $(\sqrt[3]{1000})(\sqrt{25})$

23. $-(\sqrt{49})(\sqrt[3]{8})$

24. $|6 + 4 - 18| (\sqrt[3]{216})$

25. $(\sqrt{64}) \cdot \frac{1}{2}(\sqrt[3]{64})$

26. $|-10|^2 - 5^2 - \sqrt{49}$

27. $(\sqrt[3]{343})(\sqrt{9}) + \sqrt{16}$

28. $-(\sqrt{100}) + \sqrt[3]{8} + \sqrt[3]{125}$

29. $\sqrt[3]{729} + \sqrt{81} + 4^2$

30. $7^2 + \sqrt{49} + |-7|$

31. $|-12| + |10| + 5^2 - (-5)^2$

32. $-(\sqrt{49}) + \sqrt[3]{512} - \sqrt{81}$

33. $(\sqrt{16} + \sqrt[3]{64})^2 - (6)^2$

34. $3^2 - \sqrt{81} - \sqrt[3]{8}$

35. $|-12| + |12| + 5^2 + (-5)^2$

LESSON 61 – Solving Equations with Two or More Rules

Example 1: $6x + 5 = 23$

$$
\begin{array}{rrl}
6x + & 5 = & 23 \\
\text{Step 1} \quad\quad\quad - & 5 & - 5 \quad \text{Subtract 5 from both sides.} \\
\hline
6x = & 18
\end{array}
$$

Step 2 $\quad\quad \dfrac{6x}{6} = \dfrac{18}{6}$ Divide both sides by 6.

$\quad\quad\quad\quad\quad\quad x = 3$

Example 2: $\dfrac{x}{3} - 4 = 8$

$$
\dfrac{x}{3} - 4 = 8
$$

Step 1 $\quad\quad\quad + 4 \quad\quad + 4$ Add 4 to both sides.

$$
\dfrac{x}{3} = 12
$$

Step 2 $\quad\quad \dfrac{3x}{3} = 12(3)$ Multiply both sides by 3.

$\quad\quad\quad\quad\quad x = 36$

Example 3: $4x + 6x - 8 = 22 + 30$

Step 1 $\quad 10x - 8 = 52$ Combine like terms on both sides.
Step 2 $\quad\quad\quad + 8 \quad\quad + 8$ Add 8 to both sides.

$\quad\quad\quad\quad 10x = 60$

Step 3 $\quad\quad \dfrac{10x}{10} = \dfrac{60}{10}$ Divide both sides by 10.

$\quad\quad\quad\quad\quad x = 6$

Example 4: $4x - 7 + 2 = 30 + \left(\dfrac{10}{2}\right) - 4x$

Step 1	$4x$	$-$	5	$=$	35	$- \ 4x$	Combine like terms.
Step 2		$+$	5		$+ 5$		Add 5 to both sides.

$$4x = 40 - 4x$$

Step 3 $+ \ 4x \qquad\qquad + \ 4x$ Add 4x to both sides.

$$8x = 40$$

Step 4 $\dfrac{8x}{8} = \dfrac{40}{8}$ Divide both sides by 8.

$$x = 5$$

Example 5: $\frac{1}{2}x + 3\frac{2}{3}x - 5 = 10 + 8 + 2$

Step 1 $\frac{25}{6}x \quad - \quad 5 \ = \ 20$ Combine like terms.

Step 2 $\qquad\quad + \quad 5 \qquad + 5$ Add 5 to both sides.

$$\frac{25}{6}x \ = \ 25$$

Step 3 $\frac{6}{25} \cdot \frac{25}{6}x \ = \ 25 \cdot \frac{6}{25}$ Multiply both sides by $\frac{6}{25}$, the reciprocal of $\frac{25}{6}$.

$$x \ = \ 6$$

Practice solving equations with two or more rules.

1. $5x - 1 = 9$

2. $3x + 2 = 17$

3. $4x + 7 = 23$

4. $-\frac{3}{4}x + \frac{2}{3} = \frac{1}{6}$

5. $1\frac{2}{3} + 1\frac{1}{2}x = -1\frac{1}{6}$

6. $1\frac{1}{3}x - 1\frac{1}{2} + \frac{1}{3}x = -1\frac{11}{12}$

7. $2x + 5 = 15$ a. 5   b. 20   c. 10   d. 4

8. $-6x + 2 - 3x + 2 = 3 + 2x + 4x - 4$

9. $-2x - 1 + 5x + 3 = 1 + 2x + 3x - 4$
    a. $-2\frac{1}{2}$ b. $\frac{2}{5}$ c. $2\frac{1}{2}$ d. $-\frac{2}{5}$

10. $-3x - 1 - 2x - 3 = -4 - x + 2x + 4$
    a. $\frac{2}{3}$ b. $-\frac{2}{3}$ c. $1\frac{1}{2}$ d. $-1\frac{1}{2}$

11. $3x + 1 + 5x + 5 = -4 - 4x + 3x + 2$

12. $4x + 9 = 17$ a. 2   b. 4   c. 26   d. 8

13. $2\frac{5}{6}x + 1\frac{2}{3} + \frac{1}{3}x = -2\frac{1}{12}$

14. $-1\frac{3}{4} - 2\frac{1}{2}x = -3\frac{3}{8}$

15. $-\frac{1}{6}x + \frac{1}{12} = -\frac{3}{4}$

16. $5x + 1 = 31$

17. $4x + 7 = 35$

18. $\frac{1}{3}x + \frac{1}{2} = -\frac{3}{4}$

19. $4x + 7 = 47$ a. 54   b. 40   c. 10   d. 9

20. $10x + 1 = 21$ a. 11   b. 5   c. 2   d. 22

21. $-1\frac{1}{4}x - 3\frac{1}{6} + \frac{1}{6}x = -3\frac{1}{12}$

22. $3\frac{1}{3}x - 2\frac{1}{4} + \frac{1}{3}x = -1\frac{1}{12}$

23. $8x + 5 = 61$ a. 7   b. 66   c. 8   d. 56

24. $4x + 7 = 11$ a. 18   b. 2   c. 1   d. 4

25. $-2\frac{1}{6} - 1\frac{3}{4}x = -3\frac{7}{12}$

26. $\frac{1}{3}x - \frac{1}{4} = \frac{7}{12}$

LESSON 62 – More Equations using Decimals and Fractions
One and Two Step Solutions

Example 1: $3x = 6.21$

$$\frac{3x}{3} = \frac{6.21}{3}$$ Divide both sides by 3, the coefficient of x.

$$x = 2.07$$

Example 2: $4x + \frac{1}{2} = 8\frac{3}{4} - \frac{1}{4}$

$$4x + \frac{1}{2} = 8\frac{1}{2}$$ Combine like terms.

$$-\frac{1}{2} \qquad -\frac{1}{2}$$ Subtract $\frac{1}{2}$ from both sides.

$$4x = 8$$

$$\frac{4x}{4} = \frac{8}{4}$$ Divide both sides by 4.

$$x = 2$$

Example 3: $3\frac{2}{3}y = 22 + 33$

$$3\frac{2}{3}y = 55$$ Combine like terms.

$$\frac{11}{3}y = 55$$ Convert $3\frac{2}{3}$ to an improper fraction, $\frac{11}{3}$.

$$\frac{3}{11} \cdot \frac{11}{3}y = 55 \cdot \frac{3}{11}$$ Multiply both sides by $\frac{3}{11}$, the reciprocal of $\frac{11}{3}$.

$$y = 15$$

Example 4: $0.6x + 0.7x = 5 + \frac{1}{5}$

$$1.3x = 5\frac{1}{5}$$ Combine like terms on both sides.

$$1.3x = 5.2$$ Convert $5\frac{1}{5}$ to a decimal.

$$\frac{1.3x}{1.3} = \frac{5.3}{1.3}$$ Divide both sides by 1.3, the coefficient of x.

$$x = 4$$

Example 5: $1.2x + 2.4x = 20.5 + 0.38$

$$3.6x = 20.88$$ Combine like terms on both sides.

$$\frac{3.6x}{3.6} = \frac{20.88}{3.6}$$ Divide both sides by 3.6, the coefficient of x.

$$x = 5.8$$

Practice on equations with decimal and fraction. Solve the following equations.

1. $-8h = 7.2$

2. $-4g = 3.2$

3. $-18k = -504$ a. -486 b. 28 c. -28 d. 486

4. $-15s = -345$ a. 23 b. -330 c. 330 d. -23

5. $\frac{4}{5}g - 7 = 1$

6. $\frac{2}{3}w - 5 = 7$

7. $3c + 14 = 29$

8. $7s - 18 = 3$

9. $4h + 39 = 47$

10. $0.5x - 1.6 = 1.2$

11. $0.5w - 1.2 = 1.2$

12. $1\frac{2}{3}h - 27 = 63$

13. $1\frac{4}{7}y - 22 = 55$

14. $2\frac{3}{5}p - 17 = 74$

15. $\frac{7}{9}y - 6 = 8$ a. 18 b. 19 c. $10\frac{8}{9}$ d. $2\frac{4}{7}$

16. $\frac{5}{6}y - 8 = 2$ a. $8\frac{1}{3}$ b. 12 c. 10 d. $-7\frac{1}{5}$

17. $\frac{5}{6}y - 2 = 8$ a. 12 b. 13 c. $7\frac{1}{5}$ d. $8\frac{1}{3}$

18. $0.2p + 1.2 = 1.5$

19. $4a + 38 = 70$

20. $9u - 32 = 13$

21.	$18 - N = 12$	22.	$N + 8 = 32$
23.	$9 - N = -41$	24.	$N + 12 = 47$
25.	$8 + N = 27$	26.	$64 - N = 83$
27.	$N - 23 = 15$	28.	$72 - N = 47$
29.	$N + 16 = -8$	30.	$12 - N = -23$
31.	$N + 10 = 13$	32.	$5 - N = 25$
33.	$N - 12 = 16$	34.	$23 - N = -37$
35.	$5 + N = 12$	36.	$32 - N = 63$
37.	$N + 16 = 12$	38.	$47 + N = 22$
39.	$N - 12 = 41$	40.	$55 - N = 66$
41.	$5x = 35$	42.	$\dfrac{42}{x} = 3$
43.	$64 = 4x$	44.	$12x = 132$
45.	$\dfrac{x}{4} = 21$	46.	$-8x = 48$
47.	$\dfrac{10}{-x} = 2$	48.	$3x = -18$
49.	$\dfrac{x}{5} = 17$	50.	$9x = -36$
51.	$\dfrac{x}{2} = -8$	52.	$7x = 84$
53.	$\dfrac{48}{x} = -16$	54.	$6x = -54$
55.	$\dfrac{x}{3} = 27$	56.	$4x = -84$
57.	$64 = \dfrac{x}{8}$	58.	$-3x = 12$

LESSON 63 – Solving Mixed Number Equations

Example 1: $x - \frac{1}{2} = 3\frac{2}{3}$

$$
\begin{array}{rcl}
x \;-\; \frac{1}{2} &=& 3\frac{2}{3} \\
+\; \frac{1}{2} & & +\frac{1}{2} \\
\hline
x &=& 4\frac{1}{6}
\end{array}
$$

Add $\frac{1}{2}$ to both sides.

Obtain common denominator

and then add:

Simplify:

$3\frac{2}{3} + \frac{1}{2} =$

$3\frac{4}{6} + \frac{3}{6} =$

$3\frac{7}{6}$

$3\frac{7}{6} = 4\frac{1}{6}$

Example 2: $y + 2\frac{1}{3} = 4\frac{5}{8}$

$$
\begin{array}{rcl}
y \;+\; 2\frac{1}{3} &=& 4\frac{5}{8} \\
-\; 2\frac{1}{3} & & -\, 2\frac{1}{3} \\
\hline
y &=& 2\frac{7}{24}
\end{array}
$$

Subtract $2\frac{1}{3}$ from both sides.

Obtain common denominator

and subtract:

$4\frac{5}{8} - 2\frac{1}{3} =$

$4\frac{15}{24} - 2\frac{8}{24} =$

$2\frac{7}{24}$

Example 3: $z - 1\frac{3}{5} = 7\frac{2}{3}$

$$
\begin{array}{rcl}
z \;-\; 1\frac{3}{5} &=& 7\frac{2}{3} \\
+\; 1\frac{3}{5} & & +1\frac{3}{5} \\
\hline
z &=& 9\frac{4}{15}
\end{array}
$$

Add $1\frac{3}{5}$ to both sides.

Obtain common denominator

and add:

Simplify

$7\frac{2}{3} + 1\frac{3}{5} =$

$7\frac{10}{15} + 1\frac{9}{15} =$

$8\frac{19}{15}$

$8\frac{19}{15} = 9\frac{4}{15}$

Practice problems on solving mixed number equations. Solve the following equations.

1. $x + 1\frac{1}{2} = 5\frac{3}{8}$

2. $x - 1\frac{1}{3} = 6\frac{1}{5}$

3. $x + 1\frac{3}{4} = 3\frac{1}{8}$

4. $x - 2\frac{1}{2} = 4\frac{1}{2}$

5. $x + 1\frac{3}{4} = 2\frac{3}{5}$

6. $x + 3\frac{2}{3} = 5\frac{5}{12}$

7. $x - 1\frac{5}{6} = 3\frac{7}{12}$

8. $x + 2\frac{2}{5} = 6\frac{1}{5}$

9. $x - 2\frac{1}{2} = 3\frac{3}{8}$

LESSON 64 – Distributive Property in Algebra

$$a(b + c) = ab + ac$$

$$a(b + c - d) = ab + ac - ad$$

Example 1: $2(a + b) = 2a + 2b$

Example 2: $4(x - y + z) = 4x - 4y + 4z$

Example 3: $-3(2a - 3b + 4c) = -6a + 9b - 12c$

Example 4: $6(m - 2n + p) = 6m - 12n + 6p$

Example 5: $5x(x + 2y - z) = 5x^2 + 10xy - 5xz$

Example 6: $3(x^2 + 4x - 5) = 3x^2 + 12x - 15$

Example 7: $y(x - 2y + 3z) = xy - 2y^2 + 3yz$

Example 8: $3x(2x - 5y) = 6x^2 - 15xy$

Example 9: $10(-a - 4b + 3c) = -10a - 40b + 30c$

Example 10: $-2x(x + y - 3) = -2x^2 - 2xy + 6x$

Practice on distributive property in algebra. Use the distributive property to multiply.

1. $7(7f - 6g - 3)$

2. $9(6c + 5d + 5)$

3. $8(2w - 7x + 2)$

4. $2(9s - 2t - 3)$

5. $2x(-4x + 2y - 4)$

a. $-8x^2 + 4xy - 2x$ b. $4x^2 - 2xy - 2x$
c. $-8x^2 + 4xy - 8x$ d. $-8x^2 - 8xy + 4x$

142

6. $5d(5d + 4e - 8)$

    a. $25d^2 + 20de - 40d$ b. $9d^2 - 3de + 10d$
    c. $25d^2 - 40de + 20d$ d. $25d^2 + 9de - 3d$

7. $9p(-9p + 9q + 2)$

    a. $-81p^2 + 81pq + 18p$ b. $18p^2 + 11pq$
    c. $-81p^2 + 18pq + 11p$ d. $-81p^2 + 18pq + 81p$

8. $8t(-8t + 7u + 5)$

    a. $15t^2 + 13tu$ b. $-64t^2 + 56tu + 40t$
    c. $-64t^2 + 15tu + 13t$ d. $-64t^2 + 40tu + 56t$

9. $5(7c + 2d + 4)$

10. $6(2g - 9h - 5)$

11. $2(6j - 7k + 3)$

12. $3(5f + 6g - 2)$

13. $4v(-8v + 3w - 5)$

    a. $7v^2 - vw - 4v$ b. $-32v^2 - 20vw + 12v$
    c. $-32v^2 + 7vw - 1v$ d. $-32v^2 + 12vw - 20v$

14. $6r(3r + 4s - 8)$

    a. $18r^2 + 24rs - 48r$ b. $18r^2 + 10rs - 2r$
    c. $18r^2 - 48rs + 24r$ d. $10r^2 - 2rs + 9r$

15. $6b(-4b + 6c + 2)$

    a. $-24b^2 + 12bc + 8b$ b. $12b^2 + 8bc + 2b$
    c. $-24b^2 + 12bc + 36b$ d. $-24b^2 + 36bc + 12b$

16. $8t(3t + 2u - 4)$

    a. $10t^2 + 4tu + 11t$ b. $24t^2 + 16tu - 32t$
    c. $24t^2 + 10tu + 4t$ d. $24t^2 - 32tu + 16t$

17. $3q(-6q + 5r + 6)$

    a. $-18q^2 + 15qr + 18q$ b. $-18q^2 + 18qr + 15q$
    c. $8q^2 + 9qr - 3q$ d. $-18q^2 + 8qr + 9q$

18. $5(6s - 4t + 5)$

19. $7(4m - 9n + 2)$

LESSON 65 – Solving an Expression or Equation Given Values for the Variables

Example 1: Given a = 2 and if b = 2a + 14, then what is the value of "b"?

b = 2a + 14
b = 2(2) + 14   replace "a" with 2
b = 4 + 14
b = 18

Example 2: Evaluate $\dfrac{c + d}{e}$, given c = 5, d = 22, and e = 3.

$$\frac{c + d}{e} = \frac{5 + 22}{3} = \frac{27}{3} = 9$$

Example 3: Given x = 4 and if 2x = y + 17, find "y".

2x = y + 17
2(4) = y + 17   replace "x" with 4
8 = y + 17   subtract 17 from both sides
−9 = y

Example 4: Evaluate −a(b − c), given a = −4, b = 10, and c = −5.

−a(b − c) =
−4[10 − (−5)] = −4[10 + 5] = −4[15] = −60
or, using the distributive property
−4[10 − (−5)] = −4(10) − (−4)(−5) = −40 − (+20) = −40 − 20 = −60

Example 5: Find "x" given 2x + 3y = 6y + 7 and y = 3.

2x + 3(3) = 6(3) + 7   replace "y" with 3
2x + 9 = 18 + 7
2x + 9 = 25   subtract 9 from both sides
2x = 16
x = 8

Example 6: Find "x" given 4x − 3z = −2x + 8z and z = 6

4x − 3(6) = −2x + 8(6)   replace "z" with 6
4x − 18 = −2x + 48
6x − 18 = 48   add 2x to both sides
6x = 66   add 18 to both sides
x = 11

144

Practice on solving an expression given values for the variables. Evaluate the following.

1. If $g = -4$ and $h = 4g - 2$, the "h" equals what number?

2. $\dfrac{p + q}{r}$ if $p = -1$, $q = -6$, and $r = -5$

3. $\dfrac{r + s}{t}$ if $r = -6$, $s = -7$, and $t = -7$

4. If $g = -3$ and $h = -7g - 3$, the "h" equals what number?

5. If $p = -4$ and $q = -6p + 4$, then "q" equals what number?

6. $-h(-f + g) - fg$ if $f = -7$, $g = -7$, and $h = -6$
                          a. -49 b. 49 c. -14 d. -133

7. $-e(-c + d) - cd$ if $c = -5$, $d = -8$, and $e = -5$
                          a. -23 b. 25 c. -55 d. -105

8. $\dfrac{c + d}{e}$ if $c = -6$, $d = -2$, and $e = -2$

9. $\dfrac{w + x}{y}$ if $w = -4$, $x = -5$, and $y = -5$

10. $-g(-e + f) - ef$ if $e = -1$, $f = -7$, and $g = -7$
                          a. -35 b. -63 c. -7 d. -49

11. $-f(-d + e) - de$ if $d = -6$, $e = -6$, and $f = -6$
                          a. -6 b. -36 c. 36 d. -108

In problems 12-25, solve for "x" given an equation and the value of "y".

12. $3x + 7y = 6y + 6$ $y = 6$

13. $5x + 5y = y + 5$ $y = 10$

14. $3x + 9y = 4y + 9$ $y = 3$

15. $4x + y = 3y + 8$ $y = 12$

16. $2x + 8y = 9y + 6$ $y = 2$

17. $4x + 3y = 2y + 4$ $y = 8$

18. $x + 5y = 4y + 1$ $y = 1$

19. $4x + 4y = 7y + 8$ $y = 8$

20. $3x + 9y = 2y + 9$ $y = 9$

21. $2x + 3y = 6y + 4$ $y = 4$

22. $3x + 7y = 6y + 6$ $y = 9$

23. $x + 4y = 7y + 3$ $y = 2$

24. $4x + 5y = 4y + 12$ $y = 12$

25. $5x + 2y = y + 10$ $y = 5$

26. Find "x" given $3x + 9y = 4y - 9$ and $y = 3$

27. Find "x" given $-4x + y = 3y + 8$ and $y = 12$

28. Find "x" given $2x + 8y = 9y + 6$ and $y = 2$

29. Find "x" given $2x - 3y = 4 - 6y$ and $y = -4$

30. Find "x" given $5x + 2y = -y + 10$ and $y = -5$

31. Find "x" given $4x + 3y = 2y + 4$ and $y = 8$

32. Find "x" given $-(4x + 9y) = 2y + 4$ and $y = 8$

33. Find "x" given $3x - 9y = 2y + 9$ and $y = 9$

34. Find "x" given $x + 4y = 7y + 3$ and $y = 2$

LESSON 66 – Working with Signed Numbers

Example 1: $(-6) - (7)(2) + (-5)(-8) = (-6) - 14 + 40 = -20 + 40 = 20$

Example 2: $5 + (-8)(3) - (6)(-4) = 5 + (-24) - (-24) = 5 - 24 + 24 = -19 + 24 = 5$

Example 3: $4[7 + (2)(-3) - 8(-1)] = 4[7 + (-6) - (-8)] = 4[7 - 6 + 8] = 4[9] = 36$
or, using the distributive property
$4[7 + (2)(-3) - 8(-1)] = 4(7) + 4(2)(-3) - 4(8)(-1) = 28 - 24 + 32 = 36$

Example 4: $(6) - (7)(-2) + (5)(-8) = 6 - (-14) + (-40) = 6 + 14 - 40 = -20$

Example 5: $\dfrac{9 - (-6)(3) + 6(1 + 2)}{3 + (4)(3)} = \dfrac{9 - (-18) + 6(3)}{3 + 12} = \dfrac{9 + 18 + 18}{15} = \dfrac{45}{15} = 3$

Example 6: $\dfrac{6 + (-2)^2(3) + 4(-3)}{-1 + (-2)^2} = \dfrac{6 + (4)(3) + 4(-3)}{-1 + 4} = \dfrac{6 + 12 - 12}{3} = \dfrac{6}{3} = 2$

Practice on working with signed numbers. Simplify the following.

1. $(-5) - (-6)(-7) - (+2)(-1)$

2. $(-1) - (7)(-3) - (-6)(-8)$

3. $(-8) - (-9)(8) - (-8)(-9)$

4. $\dfrac{(-14) + (-9)(+5)}{(-3) - (-8) - (+2)}$

5. $\dfrac{(-11) - (-7)(-3)}{(-2) + (-7) - (+2)}$

6. $\dfrac{(-7) + (7)(-4) - (-1)^2}{(-1) - (-2)}$

7. $\dfrac{(-4) + (2)(-8) - (-2)^2}{(-2) - (-1)}$

8. $(-1)[(-7) - (1)(-8)]$ a. −1   b. −64   c. 15   d. −48

9. $(-3)[(-5) - (-6)(6)]$ a. 126   b. −93   c. 51   d. −18

10. $(-2)[(-3) - (9)(-4)]$ a. 42   b. −96   c. 12   d. −66

11. $(-5)[(-4) - (8)(-1)]$ a. 28   b. −20   c. −12   d. −60

12. $(-3) - (-5)(1) - (-9)(-5)$

13. $(-5) - (-2)(-6) - (-1)(4)$

LESSON 67 – Division with Signed Numbers and Exponents

Example 1: $\dfrac{(4)(3) + (-5)(2)}{4} = \dfrac{12 + (-10)}{4} = \dfrac{12 - 10}{4} = \dfrac{2}{4} = \dfrac{1}{2}$

Example 2: $\dfrac{(-1)^2 + (5)(-7) + (-2)(-1)}{(2)(-2)^2} = \dfrac{(-1)(-1) + 5(-7) + (-2)(-1)}{2(-2)(-2)} =$

$\dfrac{1 - 35 + 2}{(-4)(-2)} = \dfrac{-34 + 2}{8} = \dfrac{-32}{8} = -4$

Example 3: $\dfrac{6 - (3)(-4) + (10)(-1)}{(-2)^2} = \dfrac{6 - (-12) + (-10)}{(-2)(-2)} = \dfrac{6 + 12 - 10}{4} =$

$\dfrac{18 - 10}{4} = \dfrac{8}{4} = 2$

Example 4: $\dfrac{(-8)(-1) + (-4)^2 - (1)(2)(-6)}{(-3)^2 + (-1)(-3)} = \dfrac{(8) + (-4)(-4) - (-12)}{(-3)(-3) + 3} =$

$\dfrac{8 + 16 + 12}{9 + 3} = \dfrac{36}{12} = 3$

Practice on division with signed numbers and exponents. Simplify the following.

1. $\dfrac{(-17) + (-8)(+4)}{(-4) + (-5) - (+2)}$

2. $\dfrac{(-10) + (-6)(+3)}{(-2) - (+4) - (+3)}$

3. $\dfrac{(-14) + (-2)(-5)}{(-5) + (+6) - (+5)}$

4. $\dfrac{(-13) - (-4)(+4)}{(-5) + (-8) - (+4)}$

5. $\dfrac{(-9) + (9)(-6) - (-1)2}{(-2) + (-2)}$

6. $\dfrac{(-4) - (4)(-8) - (-2)2}{(-2) - (-1)}$

7. $\dfrac{(-8) - (2)(-3) - (-2)2}{(-1) + (-2)}$

8. $\dfrac{(-8) + (8)(-7) - (-1)2}{(-2) - (-1)}$

9. $\dfrac{(-5) + (3)(-8) - (-2)2}{(-1) - (-2)}$

10. $\dfrac{(-2) - (7)(-3) - (-1)2}{(-1) + (-2)}$

11. $\dfrac{(-9) - (5)(-5) - (-2)2}{(-2) + (-2)}$

12. $\dfrac{(-15) + (-7)(-5)}{(-2) - (-6) - (+5)}$

13. $\dfrac{(-16) + (-9)(-5)}{(-5) - (-4) - (+3)}$

14. $\dfrac{(-11) - (-4)(-4)}{(-2) + (+8) - (+5)}$

15. $\dfrac{(-10) - (-6)(+4)}{(-2) + (+6) - (+2)}$

16. $\dfrac{(-12) - (-8)(-2)}{(-4) + (-2) - (+3)}$

17. $\dfrac{(-11) - (-7)(+3)}{(-5) + (-3) - (+5)}$

18. $\dfrac{(-8) - (6)(-2) - (-2)2}{(-2) + (-1)}$

19. $\dfrac{(-3) - (8)(-3) - (-1)2}{(-2) + (-2)}$

20. $\dfrac{(-5) + (9)(-9) - (-2)2}{(-1) - (-2)}$

LESSON 68 – Review on Adding Like Terms

Example 1: $6a + 3b + (-5a) - 2b =$
$6a + (-5a) + 3b - 2b =$ rearrange using the commutative property
$(6a - 5a) + (3b - 2b) =$ combine like terms
$a + b$

Example 2: $4c - 5d + 6e + 8d - 2e =$
$4c - 5d + 8d + 6e - 2e =$
$4c + (-5d + 8d) + (6e - 2e) =$
$4c - 3d + 4e$

Example 3: $4 + 8a + 6 - 5a =$
$4 + 6 + 8a - 5a =$
$(4 + 6) + (8a - 5a) =$
$10 + 3a$

Example 4: $3x + 4xy - 7y + 6xy =$
$3x + 4xy + 6xy - 7y =$
$3x + (4xy + 6xy) - 7y =$
$3x + 10xy - 7y$

Example 5: $2a^2 - 3b^2 + 4a^2b^2 + 3a^2 + 8b^2 - 6a^2b^2 =$
$(2a^2 + 3a^2) + (-3b^2 + 8b^2) + (4a^2b^2 - 6a^2b^2) =$
$5a^2 + 5b^2 - 2a^2b^2$

Example 6: $6mn^2 + 7m^2n - 5m^2n^2 + 8m^2n - 2mn^2 =$
$(6mn^2 - 2mn^2) + (7m^2n + 8m^2n) - 5m^2n^2 =$
$4mn^2 + 15m^2n - 5m^2n^2$

Practice on review of adding like terms. Simplify the following and/or select the correct answer.

1. $6x + y + 4x + 5y$

2. $5x - 9y + 9x - 4y$

3. $7x - 6y + 6x + 8y$

4. $4x - 3 - 2x + 3$

5. $6x + 3 + 5x - 1$

6. $7x + 2 - x - 2$

7. $2x + 8y - 3x - 6y$
      a. $-x + 14y$ b. $5x + 2y$ c. $5x + 14y$ d. $-x + 2y$

8. $5x + y - 2x - 2y$
      a. $7x + 3y$ b. $3x - y$ c. $7x - y$ d. $3x + 3y$

9. $7x - 7y - 7x - 3y$
      a. $14x - 4y$ b. $-4y$ c. $14x - 10y$ d. $-10y$

10. $3x - 2 + x - 4$

11. $5x + 7 - 5x - 2$

12. $7x + 4y + 4x + 6y$

13. $8x - 9y - x - y$

14. $2x + 7y - 4x - 9y$

15. $3x + 9y - 8x - 6y$

16. $3x + 6 + 6x + 4$

17. $7x + 2 - 2x + 6$

18. $7x - 8y - 4x + 9y$
      a. $3x - 17y$ b. $11x - 17y$ c. $3x + y$ d. $11x + y$

19. $6x - 9y - 5x + 3y$
      a. $x - 6y$ b. $11x - 12y$ c. $11x - 6y$ d. $x - 12y$

20. $4x + 4y - 9x + 2y$
      a. $13x + 2y$ b. $13x + 6y$ c. $-5x + 2y$ d. $-5x + 6y$

21. $9x - 5y - 3x + 4y$
      a. $6x - 9y$ b. $12x - y$ c. $6x - y$ d. $12x - 9y$

22. $3x + 7 - 4x - 1$

23. $2x + 2 - 2x + 1$

24. $8x - 3 + x + 6$

25. $7x + y + 4x - 6y$

26. $4x + 7y + 6x - y$

27. $4x - 4y - 4x + 7y$
        a. $8x - 11y$ b. $3y$ c. $8x + 3y$ d. $-11y$

28. $9x + 2y + 4x + 2y$

29. Are $4c^2d$, $3cdc$, and $4cd^2$ like terms? yes or no

30. Are $2vuv$, $3uv^2$, and $4v^2u$ like terms? yes or no

31. $8a + 6ab + 6 + 5ab + 9a + 2 + 3a$

32. $6y + 3yz + 2 + 6yz + 7y + 3 + 9y$

33. $3 + 2h^2 + 7hh + h + 4 + 6h$
        a. $14h^2 + 6h + 12$ b. $2h^2 + 7hh + 7h + 7$
        c. $16h^6 + 7$ d. $9h^2 + 7h + 7$

34. $6 + 4z^2 + 2zz + 2z + 6 + z$
        a. $9z^6 + 12$ b. $8z^2 + 2z + 36$
        c. $4z^2\,2zz + 3z + 12$ d. $6z^2 + 3z + 12$

35. $9j + 6jk + 7 + 3jk + 3j + 6 + j$

36. $8u + 4uv + 5 + 5uv + 2u + 4 + 6u$

37. Are $2b^2a$, $4bab$, and ab^2 like terms? yes or no

38. Are $5kj^2$, jkj, and $2j^2k^2$ like terms? yes or no

39. $8d + 5de + 2 + 6de + 7d + 8 + 5d$

40. $4w + 3wx + 7 + 3wx + 3w + 6 + 3w$

41. $6m + 4mn + 3 + 5mn + 2m + 4 + 2m$

42. $7y + yz + 1 + 7yz + 5y + 5 + 8y$

43. $2 + c^2 + cc + 6c + 4 + 3c$
        a. $c^2 + cc + 9c + 6$ b. $11c^6 + 6$
        c. $2c^2 + 9c + 6$ d. $c^2 + 18c + 8$

44. $1 + 4j^2 + 6jj + 2j + 9 + 8j$
    a. $4j^2 + 6jj + 10j + 9$
    b. $24j^2 + 16j + 8$
    c. $10j^2 + 10j + 10$
    d. $20j^6 + 9$

45. $8a + ab + 3 + 2ab + 4a + 7 + 7a$

46. Are $3yxy$, $2xy^2$, and $4y^2x$ like terms? yes or no

47. $7p + pq + 6 + 9pq + 5p + 2 + 3p$

48. $7 + 7f^2 + ff + 5f + 2 + f$
    a. $14f^6 + 9$
    b. $7f^2 + ff + 6f + 9$
    c. $8f^2 + 6f + 9$
    d. $7f^2 + 5f + 14$

49. Are $2n^2m$, $3nmn$, and $4mn^2$ like terms? yes or no

50. Are $3b^2c^2$, cb^2, and $5bcb$ like terms? yes or no

51. $1 + 3g^2 + 7gg + 4g + 3 + 5g$
    a. $10g^2 + 9g + 4$
    b. $3g^2 + 7gg + 9g + 4$
    c. $19g^6 + 4$
    d. $21g^2 + 20g + 3$

52. $3 + 6d^2 + 5dd + 7d + 8 + 7d$
    a. $11d^2 + 14d + 11$
    b. $25d^6 + 11$
    c. $6d^2 + 5dd + 14d + 11$
    d. $30d^2 + 49d + 24$

53. $7 + 4k^2 + 4kk + k + 4 + 4k$
    a. $13k^6 + 11$
    b. $8k^2 + 5k + 11$
    c. $4k^2 + 4kk + 5k + 11$
    d. $16k^2 + 4k + 28$

54. $6a + 7b - 8c + 4a - (3b + 4c)$

55. $15d - 5e + 8(f - 2d + 3e)$

56. $24x + 15y - 23z + 3x - (12y + 3z)$

57. $2e^2 + 5ef - 8ef^2 + 4e^2 - 2ef^2$

58. $-x^2 + 16xy + 4z^2 + 2xy - 8x^2$

59. $3(-a + b - 2c) + 4(-a^2 - b + c)$

60. $abc + a^2b^2c^2 - 8bac + 5c^2b^2a^2$

LESSON 69 – Exponents with Signed Numbers

Example 1: $(-1)^2 - (-3)^2 = (-1)(-1) - (-3)(-3) = 1 - 9 = -8$

Example 2: $-2^2 + 4^2 = -(2)(2) + (4)(4) = -4 + 16 = 12$

Example 3: $(-2)^2 + 4^2 = (-2)(-2) + (4)(4) = 4 + 16 = 20$

Example 4: $-2^2 - 4^2 = -(2)(2) - (4)(4) = -4 - 16 = -20$

Example 5: $(-2)^2 + (-4)^2 = (-2)(-2) + (-4)(-4) = 4 + 16 = 20$

Example 6: $-2^2 + (-4)^2 = -(2)(2) + (-4)(-4) = -4 + 16 = 12$

Example 7: $(-1)^3 + 2^3 - (-1)^3 = (-1)(-1)(-1) + (2)(2)(2) - (-1)(-1)(-1) =$
$-1 + 8 - (-1) = -1 + 8 + 1 = 7 + 1 = 8$

Practice problems on exponents with signed numbers. Simplify the following.

1. $-3^2 - (-2)^2$

2. $-4^2 - (-3)^2$

3. $-2^4 - (-2)^2 - 2^2$ a. −16   b. −24   c. 8   d. 16

4. $-3^2 - (-3)^3 - 2^3$ a. −26   b. −44   c. 28   d. 10

5. $-3^3 - (-3)^2 - 2^4$ a. −52   b. −34   c. 2   d. 20

6. $-3^3 - (-2)^4$

7. $-(-4)^3 - (-3)^2$

8. $-3^3 - (-2)^3 - 2^2$ a. −23   b. −39   c. 31   d. 15

9. $-2^3 - (-3)^3 - 3^3$ a. −62   b. 8   c. −46   d. −8

10. $-(-4)^2 - (-2)^2$

11. $-4^2 - (-2)^2$

12. $-3^3 - (-2)^3 - 2^3$ a. −43   b. 11   c. 27   d. −27

LESSON 70 – Variables on Both Sides of the Equation

Remember to get all unknowns to one side of the equation (preferably to the left side) and all constants to the other side before solving.

Example 1: $6x - 4x + 8 = x + 9 + 2$

$$\begin{array}{rcl} 2x \ + \ 8 & = & x \ + \ 11 \\ - \ 8 & & - \ 8 \end{array}$$ Combine like terms.
Subtract 8 from both sides.

$$\begin{array}{rcl} 2x & = & x \ + \ 3 \\ - \ x & & - \ x \end{array}$$ Subtract "x" from both sides

$$x = 3$$

Example 2: $x - 6 + 2x + 4 = 9 - 3x + 2x - 6$

$$\begin{array}{rcl} 3x \ - \ 2 & = & 3 \ - \ x \\ + \ 2 & & + \ 2 \end{array}$$ Combine like terms.
Add 2 to both sides.

$$\begin{array}{rcl} 3x & = & 5 \ - \ x \\ + \ x & & + \ x \end{array}$$ Add "x" to both sides

$$\frac{4x}{4} = \frac{5}{4}$$ Divide both sides by 4, the coefficient of "x".

$$x = \frac{5}{4}$$

Practice problems with variables on both sides of the equation. Solve the following.

1. $x - 4 + 3x + 1 = 4 - 4x + x + 1$

2. $-5x - 1 + 6x - 2 = 5 + 3x + 2x - 4$

3. $2x - 2 - 4x + 4 = -2 - 2x - 3x + 2$

4. $x - 2 + 6x + 4 = -2 + 4x + 2x - 1$

                   a. -5 b. $-\frac{1}{5}$ c. 5 d. $\frac{1}{5}$

5. $3x - 1 - 5x + 2 = -4 - 4x - x - 5$

        a. $3\frac{1}{3}$ b. $\frac{3}{10}$ c. $-\frac{3}{10}$ d. $-3\frac{1}{3}$

6. $5x + 4 - x - 2 = 4 - 4x + x + 5$

7. $3x - 2 - 6x + 3 = 3 + 2x - 3x - 4$

8. $3x - 1 - x - 1 = 5 - x - 4x + 1$

9. $-4x + 2 - 2x - 1 = 2 - 2x - 3x + 2$

10. $-3x - 5 - 5x + 1 = -2 + 2x - 4x - 1$

        a. $\frac{1}{6}$ b. $-\frac{1}{6}$ c. -6 d. 6

11. $-6x - 4 - 4x - 4 = -4 - 3x - x - 2$

        a. $-\frac{1}{3}$ b. $\frac{1}{3}$ c. 3 d. -3

12. $x - 2 - 2x - 5 = -3 + 2x + x - 5$

        a. 4 b. $\frac{1}{4}$ c. $-\frac{1}{4}$ d. -4

13. $3(x + 5) - 12 = x + 7$

14. $2.6(x - 5) = -0.4x + 13$

15. $27 - 2(x + 5) = -\frac{3}{2}(x + 10)$

16. $\frac{5}{2}(x + 6) = 4x - 9$

17. $7.4x - 23 = 5\frac{3}{5}x + 13$

18. $4N + 12 = 2(N - 10)$

19. $4x + 16 = 6x - 4$

20. $\frac{3}{2}(x + 4) = \frac{2}{3}(x - 3) - 2$

LESSON 71 – More Equations to Complete the Workbook

Practice problems on more equations. Solve the following equations.

1. $-2x + 2 = -5x - 1$

2. $3x - 2 = x + 2$

3. $-5x - 3 = -4x + 4$ a. -7 b. -8 c. $-\frac{1}{9}$ d. $-\frac{1}{7}$

4. $-3x + 16 + 3x - 7 = 27 - 6x$

5. $x + 2 = 2x - 3$ a. $\frac{1}{5}$ b. 5 c. 6 d. $-\frac{1}{3}$

6. $-4x - 1 = 5x - 5$ a. -6 b. $\frac{5}{9}$ c. $\frac{9}{4}$ d. $\frac{4}{9}$

7. $-4x + 61 - 7x - 38 = 131 + 7x$

8. $2x - 2 = 3x - 5$

9. $5x - 4 = -4x + 5$

10. $-6x - 5 = x - 3$ a. $-\frac{7}{2}$ b. $\frac{8}{5}$ c. $-\frac{3}{7}$ d. $-\frac{2}{7}$

11. $-3x - 4 = 5x - 5$ a. $\frac{1}{8}$ b. 8 c. $-\frac{9}{2}$ d. $\frac{1}{4}$

12. $x + 4 = -4x - 5$ a. $-\frac{9}{5}$ b. $-\frac{5}{9}$ c. -2 d. $\frac{1}{3}$

13. $-6x - 5 = -x - 1$ a. $\frac{6}{7}$ b. -1 c. $-\frac{5}{4}$ d. $-\frac{4}{5}$

14. $-3x + 6 - 4x - 9 = -27 - 3x$

15. $2x - 1 = 4x + 3$

16. $-2x + 1 = -5x + 5$ a. $-\frac{6}{7}$ b. $\frac{3}{4}$ c. $\frac{4}{3}$ d. $\frac{5}{3}$

17. $5x + 2 = -2x + 1$ a. $-\frac{2}{7}$ b. -7 c. $-\frac{1}{7}$ d. 1

18. $2x + 8 + 3x + 33 = -13 - 4x$

19. $5x + 3 = 6x - 1$

20. $-3x + 4 = -4x + 2$

21. $-7x - 1 + x + 4 = -5 - 2x$

22. $4x - 4 - 6x + 3 = 4 - 7x$

23. $2x + 10 + 3x - 29 = 26 - 4x$

24. $6x + 16 + 4x - 4 = -12 + 6x$

25. $-5x + 12 + x = 5x - 2 - 3x + 10$

26. $-6x - 9 + 5x = 12x + 6 - 15x + 5$

27. $-5x + 12 + 4x - x = -3x - 6 - 5x$
             a. 2   b. −5   c. −4   d. −3

28. $-x + 3x + 3x = -x - 20 + 2x$
             a. −5   b. −1   c. −6   d. −10

29. $-3x - 12 - 5x + 2x = 2x + 4x$
             a. 1   b. 2   c. −2   d. −1

30. $-4x + 8 + x - 5x = -4x - 2 + x$
             a. −2   b. 4   c. 3   d. 2

31. $-x - 2 + 6x = -2x - 6 - 11x - 3$

32. $6x - 5 + 15x = -6x - 8 + 10x - 7$

33. $-6x + 39 - x + 7 = 94 + x$

34. $2x - 8 - 4x - 2 = 35 + 7x$

35. $6x + 10 - 5x - 10 = -20 - 4x$

ANSWERS

LESSON 1
1. b 2. d 3. d 4. c 5. hundred-thousands
6. ten-millions 7. ten-billion 8. ten-thousands 9. a
10. d

LESSON 2
1. c 2. a 3. b 4. 45,170 5. 58,300 6. 46,000 7. b
8. a 9. a 10. 57,000 11. 47,300 12. 60,000
13. 4730 14. 9300 15. d 16. c 17. b 18. 68,000
19. 44,500 20. 6,000,000 21. 74,200 22. d 23. b
24. c 25. 64,400 26. 86,000

LESSON 3
1. d 2. b 3. 15 4. 32 5. 7592 6. 5142 7. 2970 8. d
9. c 10. 3667 11. 6232 12. 5484 13. 35 14. d
15. b 16. 2356

LESSON 4
1. 78,687 2. 46,257 3. d 4. b 5. b 6. 66,742 7.
1435 8. $382 \times 6 = 2292$ 9. 40,915 10. a 11. b
12. b 13. 73 R 19 14. 85 R 36 15. 8 16. 12 17. 29
18. 9 19. 43 R 54 20. 25 R 55 21. 10

22. 1048 R 3 23. $6\overline{)54}$,$54 \div 6$, $\frac{54}{6}$ 24. a 25. 15

LESSON 5

1. a, $33.25 + $13.73 = $46.98, or
$$\begin{array}{r} \$\ 33.25 \\ +\ 13.73 \\ \hline \$\ 46.98 \end{array}$$

2. b, $41.12 + $12.52 = $53.64, or
$$\begin{array}{r} \$\ 41.12 \\ +\ 12.52 \\ \hline \$\ 53.64 \end{array}$$

3. 52, 79 − 27 = 52, or
$$\begin{array}{r} 79 \\ -\ 27 \\ \hline 52 \end{array}$$

4. $0.82, $10 − $9.18 = $0.82 or
$$\begin{array}{r} \$10.00 \\ -\ 9.18 \\ \hline \$\ 0.82 \end{array}$$

5. $3.58,
$$\begin{array}{r} \$\ 5.00 \\ -\ 1.42 \\ \hline \$\ 3.58 \end{array}$$
6. 6484 candy bars,
$$\begin{array}{r} 3119 \\ +\ 3365 \\ \hline 6484 \end{array}$$

7. 17,000 seats,
$$\begin{array}{r} 24,000 \\ -\ 7,000 \\ \hline 17,000 \end{array}$$
8. a,
$$\begin{array}{r} 1433 \\ +\ 5994 \\ \hline 7427 \end{array}$$

9. 12,000 seats,
$$\begin{array}{r} 17,000 \\ -\ 5,000 \\ \hline 12,000 \end{array}$$
10. c,
$$\begin{array}{r} \$\ 64.00 \\ +\ 12.50 \\ \hline \$\ 76.50 \end{array}$$

LESSON 6
1. d 2. b 3. 8.77 4. 32.08 5. 28.35 6. b 7. c
8. 15.38 9. 8.645 10. 0.0076 11. 0.0336 12. d
13. 0.09 14. 0.084 15. 0.0516 16. 0.0792

17. 4.10,
$$\begin{array}{r} 2.29 \\ +\ 0.61 \\ \hline 2.90 \end{array} \quad \begin{array}{r} 7.00 \\ -\ 2.90 \\ \hline 4.10 \end{array}$$

LESSON 7
1. 0.21 2. 0.7 3. 0.06 4. 0.41 5. 0.71 6. a 7. d

8. 0.42 9. 0.71 10. 0.21 11. 0.41 12. 0.32 13. b
14. d 15. c 16. a 17. 0.002 18. 0.06 19. 2.3
20. 0.092 21. 0.375 22. 2.54 23. 2.664 24. 5.06
25. 1.25

LESSON 8
1. c 2. b 3. $\frac{83}{100}$ 4. $\frac{27}{100}$ 5. 0.50 6. $\frac{11}{100}$ 7. $\frac{47}{100}$ 8. d 9. b

10. c 11. d 12. c 13. d 14. b 15. $\frac{79}{100}$ 16. $\frac{29}{100}$ 17. a

18. d 19. c 20. b 21. d 22. a 23. c 24. $\frac{73}{100}$ 25. $\frac{37}{100}$

26. 0.70

LESSON 9
1. no 2. a 3. c 4. c 5. b 6. d 7. a 8. no 9. yes
10. no 11. no 12. yes 13. d 14. b 15. a 16. c
17. no 18. no 19. no 20. yes 21. yes 22. no
23. yes 24. yes 25. no 26. yes 27. yes 28. yes
29. no 30. yes

LESSON 10
1. 31 2. 83 3. c 4. b 5. 73 6. 47 7. b 8. a 9. 18,
21, 26, 10 10. 12, 16, 18, 6 11. d 12. d 13. b
14. 22, 26, 32, 12 15. a 16. 4, 6, 8, 9, 10, 12, 14,
15, 16, 18 17. 20, 21, 22, 24, 25, 26, 27, 28, 30
18. 23, 29, 31, 37, 41, 43, 47 19. 8, 35, 77, 54,
33, 57 20. 62, 63, 64, 65, 66, 68, 69 21. 4; 13,
17, 19, 23 22. False 23. 4 24. 4, 6, 8, 9

LESSON 11
1. a 2. d 3. 2•2•5•5•7 4. 3•5•5•7 5. 2•2•3•3•5•7
6. c 7. b 8. a 9. 2•2•2•3•5•5•7 10. 2•2•3•3•5•5•11
11. 2•2•3•5•7 12. d 13. b 14. b 15. 2•2•2•5•11
16. 2•2•2•3•3•5•5•11 17. 2•2•3•3•5•5•7 18. b
19. d 20. b

LESSON 12
1. a 2. c 3. a 4. a 5. a 6. b 7. c 8. 10, 20, 30, 40,
50, 60 9. 12, 24, 36, 48, 60 10. 11, 22, 33, 44,
55, 66, 77, 88, 99

LESSON 13
1. b 2. a 3. 8 4. 12 5. 30 6. 24 7. 72 8. 10 9. 18
10. c 11. 24 12. 12 13. 35 14. 60 15. 12 16. 10
17. c 18. 60 19. 30 20. 45 21. 24 22. 120 23. 24
24. 50 25. 72

LESSON 14
1. b 2. b 3. 8 4. 25 5. 16 6. 64 7. 125 8. 49 9. d
10. a 11. c 12. 36 13. 64 14. 343 15. 100 16. 81
17. 256 18. 64 19. 512 20. 729 21. 1000 22. 625
23. 121 24. 225 25. 144 26. 169 27. 32 28. -343
29. 216 30. $4x^2$ 31. $-125x^3$ 32. $100a^2$ 33. $2a^3y^3$

LESSON 15
1. 8 2. 5 3. 9 4. 16 5. 6 6. 16 7. 12 8. 16 9. 38
10. 42 11. 22 12. 12 13. 6 14. 6 15. 2 16. 24
17. 24 18. 12 19. 6 20. 30

LESSON 16
1. a 2. c 3. c 4. b 5. c 6. b 7. b 8. a 9. c 10. b

A - 1

11. c **12.** $\frac{3}{5}$ **13.** $\frac{5}{6}$ **14.** $\frac{2}{5}$ **15.** $\frac{3}{4}$ **16.** b **17.** c **18.** d **19.** b
20. $\frac{13}{20}$ **21.** $\frac{1}{3}$ **22.** $\frac{2}{5}$ **23.** $\frac{2}{3}$ **24.** $\frac{7}{8}$ **25.** $\frac{3}{10}$ **26.** $\frac{3}{5}$ **27.** $\frac{1}{3}$
28. $\frac{5}{12}$ **29.** $\frac{1}{6}$ **30.** $\frac{7}{15}$ **31.** $\frac{7}{8}$ **32.** $\frac{2}{3}$ **33.** $\frac{1}{4}$ **34.** $\frac{2}{5}$ **35.** $\frac{1}{2}$
36. $\frac{3}{8}$ **37.** $\frac{2}{5}$ **38.** $\frac{2}{3}$ **39.** $\frac{7}{10}$ **40.** $\frac{1}{4}$

LESSON 17
1. a. $-\frac{1}{9}$, **b.** 9 **2.** Zero does not have a reciprocal because division by zero is undefined. **3.** b **4.** c
5. b **6. a.** 8, **b.** $-\frac{1}{8}$ **7. a.** -2, **b.** $\frac{1}{2}$ **8.** d **9.** c **10.** a
11. b **12.** b

LESSON 18
1. $\frac{8}{35}$ **2.** $\frac{4}{63}$ **3.** d **4.** d **5.** $\frac{1}{6}$ **6.** $\frac{1}{3}$ **7.** $\frac{4}{55}$ **8.** $\frac{4}{15}$ **9.** $\frac{1}{4}$ **10.** $\frac{24}{77}$
11. $1\frac{1}{3}$ **12.** a **13.** $\frac{1}{5}$ **14.** $\frac{4}{21}$ **15.** $\frac{8}{55}$ **16.** 4 **17.** a **18.** a
19. $2\frac{1}{4}$ **20.** b **21.** $2\frac{1}{2}$ **22.** $3\frac{2}{3}$ **23.** $5\frac{1}{2}$ **24.** $1\frac{1}{2}$ **25.** $\frac{1}{3}$

LESSON 19
1. b **2.** d **3.** b **4.** b **5.** b **6.** b **7.** c **8.** a **9.** a **10.** a
11. d **12.** b

LESSON 20
1. d **2.** c **3.** $\frac{14}{5}$ **4.** $\frac{3}{2}$ **5.** a **6.** $\frac{5}{2}$ **7.** $\frac{11}{4}$ **8.** $\frac{17}{5}$ **9.** a **10.** b
11. d **12.** $\frac{21}{5}$ **13.** $\frac{51}{7}$ **14.** $\frac{35}{4}$ **15.** $\frac{35}{8}$ **16.** $\frac{20}{3}$ **17.** $\frac{29}{5}$
18. $\frac{42}{11}$

LESSON 21
1. d **2.** $\frac{16}{27}$ **3.** $\frac{1}{2}$ **4.** $\frac{10}{21}$ **5.** $\frac{4}{15}$ **6.** $\frac{1}{3}$ **7.** $\frac{28}{45}$ **8.** $\frac{27}{55}$ **9.** $\frac{10}{33}$ **10.** d
11. c **12.** $\frac{1}{3}$ **13.** $\frac{16}{45}$ **14.** $\frac{1}{4}$ **15.** $\frac{16}{45}$ **16.** $\frac{20}{33}$ **17.** $\frac{8}{27}$ **18.** $\frac{8}{35}$
19. $\frac{4}{27}$ **20.** a **21.** $\frac{40}{77}$ **22.** $\frac{8}{15}$ **23.** $\frac{8}{9}$ **24.** 4 **25.** $\frac{3}{10}$ **26.** $\frac{3}{20}$
27. $\frac{1}{15}$ **28.** $\frac{3}{28}$ **29.** $\frac{2}{3}$ **30.** $\frac{1}{4}$ **31.** $\frac{2}{3}$ **32.** $\frac{1}{8}$ **33.** $\frac{1}{3}$ **34.** 8
35. $1\frac{1}{7}$

LESSON 22
1. $\frac{2}{15}$ **2.** $3\frac{1}{3}$ **3.** $\frac{2}{3}$ **4.** 50 **5.** $2\frac{1}{3}$ **6.** $1\frac{1}{8}$ **7.** $\frac{6}{7}$ **8.** $\frac{3}{25}$ **9.** $3\frac{8}{9}$
10. $1\frac{7}{9}$ **11.** $\frac{5}{12}$ **12.** $1\frac{17}{64}$ **13.** 1 **14.** $\frac{20}{63}$ **15.** $\frac{5}{6}$ **16.** $1\frac{1}{5}$
17. 21 **18.** $1\frac{3}{35}$ **19.** $2\frac{4}{13}$ **20.** $1\frac{2}{15}$

LESSON 23
1. a **2.** a **3.** c **4.** d **5.** $\frac{25}{33}$ **6.** $2\frac{1}{13}$ **7.** $\frac{28}{45}$ **8.** $15\frac{1}{2}$ **9.** $\frac{1}{48}$
10. $\frac{3}{10}$ **11.** $37\frac{5}{7}$ **12.** $121\frac{1}{42}$ **13.** $31\frac{13}{27}$ **14.** $42\frac{9}{10}$ **15.** $\frac{3}{4}$
16. $\frac{44}{133}$ **17.** b **18.** c

LESSON 24
1. $13\frac{4}{15}$ **2.** $12\frac{19}{30}$ **3.** $12\frac{5}{12}$ **4.** $3\frac{17}{24}$ **5.** $15\frac{14}{15}$ **6.** c **7.** c **8.** a
9. $15\frac{5}{12}$ **10.** $7\frac{4}{15}$ **11.** c **12.** d **13.** $9\frac{3}{20}$ **14.** $10\frac{39}{40}$
15. $17\frac{7}{15}$ **16.** $7\frac{3}{14}$ **17.** $12\frac{1}{6}$ **18.** $15\frac{33}{35}$

LESSON 25
1. $18\frac{17}{20}$ **2.** $12\frac{7}{18}$ **3.** c **4.** b **5.** c **6.** a **7.** $18\frac{9}{20}$ **8.** $17\frac{5}{18}$
9. $19\frac{4}{45}$ **10.** $24\frac{1}{48}$ **11.** $16\frac{5}{18}$ **12.** $1\frac{5}{14}$ **13.** $1\frac{3}{8}$ **14.** $16\frac{5}{12}$
15. $9\frac{40}{63}$ **16.** $17\frac{11}{18}$ **17.** a **18.** a **19.** b **20.** $13\frac{2}{63}$
21. $1\frac{5}{14}$ **22.** $1\frac{5}{12}$ **23.** $15\frac{43}{60}$ **24.** $32\frac{49}{90}$ **25.** $21\frac{11}{12}$
26. $14\frac{33}{40}$ **27.** $22\frac{107}{120}$ **28.** $13\frac{27}{40}$ **29.** $16\frac{19}{40}$ **30.** $59\frac{1}{20}$
31. 40 **32.** $39\frac{7}{8}$

LESSON 26
1. a **2.** c **3.** b **4.** d **5.** a **6.** a **7.** $5\frac{7}{12}$ **8.** $6\frac{15}{56}$ **9.** $13\frac{14}{45}$
10. $4\frac{5}{12}$ **11.** $7\frac{7}{24}$ **12.** $11\frac{14}{45}$ **13.** $1\frac{8}{15}$ **14.** $4\frac{7}{12}$ **15.** $7\frac{19}{30}$
16. $1\frac{29}{40}$ **17.** $4\frac{3}{5}$ **18.** $14\frac{2}{9}$

LESSON 27
1. b **2.** a **3.** $4\frac{1}{3}$ **4.** $4\frac{7}{12}$ **5.** $\frac{4}{9}$ **6.** $2\frac{1}{24}$ **7.** $15\frac{8}{35}$ **8.** $14\frac{1}{20}$
9. $2\frac{11}{30}$ **10.** $5\frac{25}{28}$ **11.** $5\frac{2}{3}$ **12.** $4\frac{1}{3}$ **13.** c **14.** b **15.** a **16.** d
17. b **18.** d **19.** c **20.** $2\frac{1}{4}$ **21.** $8\frac{1}{5}$ **22.** $27\frac{17}{45}$ **23.** $9\frac{11}{12}$
24. $25\frac{23}{30}$ **25.** $\frac{13}{15}$ **26.** $26\frac{5}{6}$ **27.** $24\frac{29}{56}$ **28.** $6\frac{17}{18}$ **29.** $2\frac{1}{4}$
30. $9\frac{14}{15}$ **31.** $17\frac{37}{120}$ **32.** $3\frac{7}{24}$ **33.** $11\frac{11}{30}$ **34.** $11\frac{37}{60}$
35. $10\frac{7}{12}$

LESSON 28
1. $2\frac{9}{10}$ **2.** $1\frac{2}{3}$ **3.** 5 **4.** $4\frac{7}{8}$ **5.** $5\frac{2}{3}$ **6.** $1\frac{11}{12}$ **7.** $4\frac{7}{10}$ **8.** 9 **9.** 6
10. $3\frac{4}{5}$ **11.** $2\frac{14}{15}$ **12.** $\frac{3}{20}$ **13.** $4\frac{2}{3}$ **14.** $7\frac{4}{5}$ **15.** $1\frac{4}{5}$ **16.** 5
17. $3\frac{1}{2}$ **18.** $4\frac{7}{8}$ **19.** $1\frac{13}{24}$ **20.** $1\frac{14}{15}$

LESSON 29
1. 25% **2.** 62.5% **3.** 68% **4.** 106% **5.** 225%
6. 550% **7.** 30% **8.** 85% **9.** 420% **10.** 337.5%
11. 48% **12.** 375% **13.** 412% **14.** 1250% **15.** 60%
16. 75% **17.** $\frac{18}{25}$ **18.** $1\frac{13}{20}$ **19.** $\frac{7}{50}$ **20.** $\frac{2}{5}$ **21.** $1\frac{3}{4}$ **22.** $\frac{19}{100}$
23. $\frac{11}{200}$ **24.** $4\frac{1}{2}$ **25.** $\frac{3}{5}$ **26.** $\frac{19}{50}$ **27.** $\frac{47}{50}$ **28.** $1\frac{8}{25}$ **29.** $\frac{5}{6}$
30. $2\frac{1}{5}$ **31.** $\frac{11}{20}$ **32.** $\frac{43}{50}$ **33.** 35% empty **34.** 75%
35. 12 games **36.** 80 women

LESSON 30
1. b **2.** 10% **3.** 20% **4.** 10% **5.** 20% **6.** 10% **7.** d
8. c **9.** c **10.** a **11.** a **12.** 20% **13.** 35% **14.** $33\frac{1}{3}$%
15. 56% **16.** 125% **17.** 66% **18.** 20%

LESSON 31
1. $\frac{1}{2}$ **2.** $\frac{1}{5}$ **3.** $\frac{1}{4}$ **4.** a **5.** b **6.** a **7.** $\frac{1}{5}$ **8.** b **9.** $\frac{1}{4}$ **10.** d **11.** 2
12. 4 **13.** c **14.** $\frac{1}{5}$ **15.** $\frac{3}{8}$ **16.** $\frac{1}{2}$ **17.** $\frac{1}{2}$ **18.** d **19.** 4 **20.** 3
21. d

LESSON 32
1. a 2. 0.08 3. 0.94 4. 0.09 5. c 6. c 7. 0.05
8. 0.06 9. 0.41 10. 0.86 11. 0.03 12. b 13. 0.35
14. 0.18 15. 0.55 16. 0.64 17. b 18. c

LESSON 33
1. $\frac{6}{5}$ = 1$\frac{1}{5}$, 120% 2. 0.28, 28% 3. $\frac{23}{10}$ = 2$\frac{3}{10}$, 2.3

4. 0.55, 55% 5. $\frac{9}{4}$ = 2$\frac{1}{4}$, 2.25 6. $\frac{9}{20}$, 45% 7. d 8. d

9. a 10. c 11. $\frac{3}{20}$, 0.15 12. $\frac{17}{10}$ = 1$\frac{7}{10}$, 170% 13. 0.2,

20% 14. $\frac{9}{5}$ = 1$\frac{4}{5}$, 1.8 15. c

LESSON 34
1. b 2. $\frac{21}{44}$ 3. 16$\frac{2}{7}$ 4. 8$\frac{2}{5}$ 5. 12$\frac{9}{23}$ 6. $\frac{11}{14}$ 7. 2$\frac{4}{15}$ 8. a 9. b

10. $\frac{35}{52}$ 11. a 12. d 13. a 14. $\frac{5}{7}$ 15. 4$\frac{3}{13}$ 16. $\frac{26}{121}$

17. 8$\frac{2}{3}$ 18. 10 19. 13 20. 5$\frac{5}{13}$

LESSON 35
1. $\frac{1}{8}$ = 0.125 2. $\frac{3}{4}$ = 0.75 3. 588 4. 120 5. b 6. a

7. b 8. 320 9. 465 10. $\frac{1}{4}$ = 0.25 11. $\frac{1}{4}$ = 0.25

12. 1290 13. 300 14. b 15. d

LESSON 36
1. 285 2. 268 3. 225 4. 180% 5. 100 6. 153
7. 300% 8. 300 9. b 10. d 11. 119 12. 225%
13. 300 14. 210 15. 184 16. 180 17. d 18. 24
19. 200% 20. 250

LESSON 37
1. 5% 2. 2% 3. 2% 4. 360 5. 195 6. 600 7. a 8. a
9. a 10. 375 11. 2% 12. 480 13. b 14. c 15. 1440
16. 102 17. a 18. 70 19. 57 20. d 21. 28 22. 136
23. a 24. c 25. 56 26. 117 27. b 28. a 29. 48
30. 52 31. 50 32. 64 33. 140 34. 35 35. 150
36. 400 37. 58$\frac{1}{3}$% 38. 35 39. 57.6 40. 333$\frac{1}{3}$%

LESSON 38
1. 275% 2. 200% 3. 6800 4. 450 5. 2000 6. 200%
7. $104,760 8. $119.70 9. 40¢ per pound
10. $138.60 11. $160,160 12. $112.10
13. $147,600 14. $7.20 15. 60¢ per pound

LESSON 39
1. $\frac{8}{15}$ 2. $\frac{8}{49}$ 3. c 4. a 5. a 6. b 7. $\frac{15}{28}$ 8. $\frac{8}{7}$ 9. b 10. $\frac{32}{63}$

11. $\frac{36}{65}$ 12. 21$\frac{7}{24}$ 13. 10$\frac{1}{5}$ 14. 8$\frac{32}{71}$ 15. 34$\frac{1}{8}$

LESSON 40
1. $4.32 2. $6.10 3. b 4. d 5. $4.39 6. $5.22 7. a
8. $5.66 9. b 10. c 11. $2.51 12. $4.75 13. $3.03
14. $42 15. 2 16. $5.75 17. $12.06, yes, 1 more
pair 18. $6.60

LESSON 41
1. 22 gallons 2. 7 gallons 3. 23 pounds 4. a 5. a
6. 25 pounds 7. 29 gallons 8. 26 gallons 9. b
10. b 11. 13 gallons 12. 25 pounds 13. c 14. b
15. c 16. 29 pounds 17. 17 gallons 18. b 19. a

20. 70 mpg 21. 15 mph 22. 80 miles 23. 30
gallons 24. c

LESSON 42
1. 27 mph 2. 24 mph 3. 3$\frac{3}{5}$ mph 4. 9$\frac{1}{3}$ mph 5. 24
km/hr 6. 54 mph 7. 28 mph 8. 40 mph 9. 55 mph
10. 350 miles 11. 7$\frac{1}{2}$ hrs 12. 50 mph 13. 60 mph
14. 520 miles 15. 12 hr

LESSON 43
1. 11 mph 2. 4$\frac{2}{3}$ mph 3. 2$\frac{2}{3}$ mph 4. 5 cookies 5. c
6. d 7. 7 cookies 8. 6 questions 9. d 10. 6$\frac{2}{5}$ mph
11. 5$\frac{1}{3}$ mph 12. 10$\frac{2}{3}$ mph 13. 36 cookies 14. 15
men 15. 8 mph

LESSON 44
1. $70.00 2. 31.5 feet 3. $39.90 4. $156 5. a
6. 247 doves 7. 1$\frac{2}{3}$ cups of raisins 8. c 9. 11 bad
guys 10. 40 km/hr 11. 4$\frac{4}{5}$ mph 12. $2.40
13. $0.07 14. $3.51 15. d 16. a 17. 456 dwarfs
18. 208 sailors 19. b 20. d 21. 76 dwarfs 22. 51
rascals 23. 280 pirates 24. 11 good guys 25. c
26. d 27. d 28. b 29. 40 rascals 30. 144 dwarfs
31. a 32. d 33. 119 fairies 34. 24 bad guys

LESSON 45
1. $1.47 2. $5.22 3. 128% 4. 176% 5. 130% 6. d
7. $3.01 8. b 9. $3.35 10. 134% 11. $54.25
12. 62.5% 13. $12,000

LESSON 46
1. 560 frogs 2. 272 acres 3. 594 hectares 4. a
5. c 6. 195 acres 7. 1284 toads 8. 2808 frogs
9. 544 acres 10. 3664 frogs 11. 1010 toads
12. 750 frogs 13. d 14. a 15. a 16. b 17. 3276
toads 18. 460 frogs 19. 420 acres 20. a 21. d

LESSON 47
1. a. 0.031 b. 0.00005 c. 0.1025 d. 0.009
2. a. 0.049 b. 0.00017 c. 0.3145 d. 0.005
3. a. 0.072 b. 0.00064 c. 0.932 d. 0.001
4. a. 0.021 b. 0.00002 c. 0.1655 d. 0.0065
5. a. 0.0341 b. 0.995 c. 0.006 6. a. 0.0508
b. 0.0375 c. 0.0005 7. a. 0.0829 b. 0.8125
c. 0.0045 8. a. 3.2% b. 0.01% c. 14.55% d. 0.5%
9. a. 7.1% b. 0.032% c. 49.15% d. 0.3%
10. a. 80% b. 344% c. 102% d. 3.6%
11. a. 900%, b. 65% c. 4% d. 220% e. 41$\frac{2}{3}$%
f. 62% 12. a. 0.15 b. 0.18 c. 0.64 d. 1.75
13. a. 49% b. 123% c. 7.6% or 7$\frac{3}{5}$% d. 456%
e. 98.7% or 98$\frac{7}{10}$% 14. a. 87% b. 4% c. 137%
d. 6.4% or 6$\frac{2}{5}$% 15. a. $\frac{341}{10000}$ b. $\frac{199}{200}$ c. $\frac{3}{500}$

16. a. $\frac{127}{2500}$ **b.** $\frac{3}{80}$ **c.** $\frac{1}{2000}$ **17. a.** $\frac{829}{10000}$ **b.** $\frac{13}{16}$ **c.** $\frac{9}{2000}$

18. a. 15% **b.** 18% **c.** 64% **19. a.** 175% **b.** 350%

c. 1525% **20. a.** $237\frac{1}{2}$% **b.** $512\frac{1}{2}$% **c.** 835%

LESSON 48
1. +2 **2.** -20 **3.** -12 **4.** +2 **5.** +5 **6.** b **7.** c **8.** b **9.** b
10. +10 **11.** +10 **12.** +2 **13.** -3 **14.** +1 **15.** -10
16. d **17.** d **18.** b **19.** b **20.** c **21.** a **22.** c **23.** +10
24. +2 **25.** -12 **26.** d **27.** c **28.** -15 **29.** -2 **30.** -2
31. 2 **32.** 10 **33.** 19

LESSON 49
1. a. 4 **b.** -27 **c.** 4 **d.** -5 **2. a.** 42 **b.** -28 **c.** 5 **d.** -4
3. a. -1120 **b.** 550 **c.** 10 **d.** -50 **4. a.** -570 **b.** 980
c. 9 **d.** -70 **5.** b **6.** a **7.** b **8.** c **9.** a **10.** d **11. a.** -640
b. 850 **c.** 9 **d.** -30 **12. a.** 20 **b.** -10 **c.** 4 **d.** -5
13. a. 24 **b.** -16 **c.** 3 **d.** -4 **14.** c **15.** c **16.** a **17.** d
18. b **19. a.** -1280 **b.** 300 **c.** 8 **d.** -20 **20. a.** 72
b. -24 **c.** 4 **d.** -3

LESSON 50
1. a **2.** 15 **3.** 9 **4.** 9 **5.** a **6.** b **7.** c **8.** d **9.** c **10.** 19
11. d **12.** b **13.** c **14.** d **15.** d **16.** d **17.** -16 **18.** 6
19. -208 **20.** 61 **21.** b **22.** d **23.** -4 **24.** a **25.** a
26. 2 **27.** -148 **28.** -19 **29.** b **30.** 28 **31.** 31 **32.** 14
33. d **34.** b **35.** 22 **36.** 14 **37.** -18 **38.** 552 **39.** 86
40. b **41.** 42 **42.** 148 **43.** b **44.** 114 **45.** 23 **46.** -18
47. 20 **48.** 11

LESSON 51
1. d **2.** a **3.** -16 **4.** -20 **5.** -25 **6.** 19 **7.** -9 **8.** -30
9. -42 **10.** c **11.** d **12.** 0 **13.** 14 **14.** -13 **15.** 45
16. c **17.** a **18.** b **19.** 24 **20.** -1 **21.** 23 **22.** -15
23. -11 **24.** 7 **25.** -35 **26.** 24 **27.** -30 **28.** -10
29. 90 **30.** -23 **31.** -90 **32.** -360 **33.** 0 **34.** 14
35. 88

LESSON 52
1. N = 26 **2.** N = 34 **3.** d **4.** a **5.** a **6.** r = 24
7. t = 16 **8.** c = 14 **9.** N = 28 **10.** N = 26 **11.** g = 25
12. a **13.** N = 27 **14.** N = 31 **15.** N = 33 **16.** d
17. a **18.** N = 37 **19.** b **20.** b **21.** j = 22 **22.** d = 20
23. N = 27 **24.** N = 37 **25.** N = -20 **26.** a = 34
27. b = 42 **28.** x = -19 **29.** c = -18 **30.** b = -18
31. D = 26 **32.** x = -20 **33.** N = -15 **34.** x = 7
35. a = -8

LESSON 53
1. c **2.** b **3.** m = 7 **4.** c = 3 **5.** p = 15 **6.** s = 8
7. z = 40 **8.** m = 5 **9.** b = 3 **10.** r = 9 **11.** w = 3
12. f = 4 **13.** a **14.** d **15.** n = 60 **16.** c = 30
17. N = 35 **18.** N = 42 **19.** N = 15 **20.** j = 80
21. c **22.** x = 99 **23.** N = 15 **24.** d **25.** d = 33
26. p = 18 **27.** c **28.** c **29.** u = 40 **30.** x = 21
31. a = 60 **32.** x = 32 **33.** N = 160 **34.** x = 15
35. x = -6 **36.** x = 4.5 **37.** c = 56 **38.** x = 6

LESSON 54
1. b **2.** 22 **3.** 31 **4.** 22 **5.** 50 **6.** 13 **7.** 10 **8.** 16 **9.** 0
10. 25 **11.** 78 **12.** c **13.** 20 **14.** 171 **15.** 107

LESSON 55
1. b **2.** b **3.** 18 **4.** 34 **5.** b **6.** d **7.** 77 **8.** -26 **9.** -99
10. 427 **11.** 3 **12.** $\frac{3}{17}$ **13.** $\frac{15}{11} = 1\frac{4}{11}$ **14.** -13 **15.** 88
16. 21 **17.** d **18.** a **19.** 34 **20.** 36 **21.** d **22.** -24
23. -19

LESSON 56
1. 3 **2.** 6 **3.** 7 **4.** 12 **5.** 10 **6.** 9 **7.** 11 **8.** 6 **9.** 15
10. 13 **11.** 24 **12.** 8 **13.** 21 **14.** 14 **15.** 24 **16.** 4
17. 7 **18.** 35 **19.** 4 **20.** 10 **21.** 2 **22.** 21 **23.** 15
24. 72 **25.** 80

LESSON 57
1. 3 **2.** 5 **3.** 19 **4.** 8 **5.** 6 **6.** 21 **7.** 30 **8.** 18 **9.** 12
10. 25 **11.** 60 **12.** 20 **13.** 42 **14.** 15 **15.** -2 **16.** 10
17. 6 **18.** 7 **19.** 8 **20.** 6 **21.** 25 **22.** 1 **23.** 12 **24.** 42
25. 20 **26.** $25x^3$ **27.** 11 **28.** 56 **29.** 27 **30.** 81 **31.** 4
32. 1.6 **33.** 25

LESSON 58
1. b **2.** a **3.** b **4.** 19 **5.** -39 **6.** -49 **7.** 6 **8.** 30 **9.** -24
10. -7 **11.** 20 **12.** -31 **13.** -3 **14.** 11 **15.** -26 **16.** 21

LESSON 59
1. -8 **2.** 16 **3.** -3 **4.** 16 **5.** d **6.** b **7.** c **8.** 4 **9.** b **10.** a
11. b **12.** d **13.** 2 **14.** 3 **15.** -3 **16.** -17 **17.** 23 **18.** b
19. 19 **20.** -4 **21.** 2 **22.** 42 **23.** 8 **24.** 84 **25.** 24
26. 14 **27.** 4 **28.** 6 **29.** 4 **30.** 22

LESSON 60
1. 3 **2.** -20 **3.** 16 **4.** 23 **5.** -4 **6.** 2 **7.** 3 **8.** 8 **9.** 12
10. b **11.** 11 **12.** -26 **13.** d **14.** c **15.** -10 **16.** 16
17. 0 **18.** -13 **19.** -16 **20.** c **21.** d **22.** d **23.** -14
24. -20 **25.** -8 **26.** 23 **27.** -1

A SHORT REVIEW
1. 78 **2.** 144 **3.** 10 **4.** 5 **5.** -13 **6.** -12 **7.** 84 **8.** -4
9. -1 **10.** 27 **11.** 22 **12.** 6 **13.** 11 **14.** -8 **15.** 21
16. 6 **17.** 72 **18.** 4 **19.** 15 **20.** 49 **21.** 3 **22.** 50
23. -14 **24.** 48 **25.** 16 **26.** 68 **27.** 25 **28.** -3 **29.** 34
30. 63 **31.** 22 **32.** -8 **33.** 28 **34.** -2 **35.** 74

LESSON 61
1. 2 **2.** 5 **3.** 4 **4.** $\frac{2}{3}$ **5.** $-\frac{17}{9} = -1\frac{8}{9}$ **6.** $-\frac{1}{4}$ **7.** a **8.** $\frac{1}{3}$ **9.** c
10. b **11.** $-\frac{8}{9}$ **12.** a **13.** $-\frac{45}{38} = -1\frac{7}{38}$ **14.** $\frac{13}{20}$ **15.** 5
16. 6 **17.** 7 **18.** $-\frac{15}{4} = -3\frac{3}{4}$ **19.** c **20.** c **21.** $-\frac{1}{13}$ **22.** $\frac{7}{22}$
23. a **24.** c **25.** $\frac{17}{21}$ **26.** $\frac{5}{2} = 2\frac{1}{2}$

LESSON 62
1. h = -0.9 **2.** g = -0.8 **3.** b **4.** a **5.** g = 10 **6.** w = 18
7. c = 5 **8.** s = 3 **9.** h = 2 **10.** x = 5.6 **11.** w = 4.8
12. h = 54 **13.** y = 49 **14.** p = 35 **15.** a **16.** b **17.** a
18. p = 1.5 **19.** a = 8 **20.** u = 5 **21.** N = 6
22. N = 24 **23.** N = 50 **24.** N = 35 **25.** N =19
26. N = -19 **27.** N = 38 **28.** N = 25 **29.** N = -24
30. N = 35 **31.** N = 3 **32.** N = -20 **33.** N = 28
34. N = 60 **35.** N = 7 **36.** N = -31 **37.** N = -4
38. N = -25 **39.** N = 53 **40.** N = -11 **41.** x = 7
42. x = 14 **43.** x = 16 **44.** x = 11 **45.** x = 84
46. x = -6 **47.** x = -5 **48.** x = -6 **49.** x = 85

50. x = -4 **51.** x = -16 **52.** x = 12 **53.** x = -3
54. x = -9 **55.** x = 81 **56.** x = -21 **57.** x = 512
58. x = -4

LESSON 63

1. $3\frac{7}{8}$ **2.** $7\frac{8}{15}$ **3.** $1\frac{3}{8}$ **4.** 7 **5.** $\frac{17}{20}$ **6.** $1\frac{3}{4}$ **7.** $5\frac{5}{12}$ **8.** $3\frac{4}{5}$ **9.** $5\frac{7}{8}$

LESSON 64

1. 49f - 42g - 21 **2.** 54c + 45d + 45
3. 16w - 56x + 16 **4.** 18s - 4t - 6 **5.** c **6.** a **7.** a **8.** b
9. 35c + 10d + 20 **10.** 12g - 54h - 30
11. 12j - 14k + 6 **12.** 15f + 18g - 6 **13.** d **14.** a
15. d **16.** b **17.** a **18.** 30s - 20t + 25
19. 28m - 63n + 14

LESSON 65

1. -18 **2.** $1\frac{2}{5}$ **3.** $1\frac{6}{7}$ **4.** 18 **5.** 28 **6.** a **7.** c **8.** 4 **9.** $1\frac{4}{5}$
10. d **11.** b **12.** 0 **13.** -7 **14.** -2 **15.** 8 **16.** 4 **17.** -1
18. 0 **19.** 8 **20.** -18 **21.** 8 **22.** -1 **23.** 9 **24.** 0 **25.** 1
26. -8 **27.** -8 **28.** 4 **29.** 8 **30.** 5 **31.** -1 **32.** -23
33. 36 **34.** 9

LESSON 66

1. -45 **2.** -28 **3.** -8 **4.** $-19\frac{2}{3}$ **5.** $2\frac{10}{11}$ **6.** -36 **7.** 24 **8.** a
9. b **10.** d **11.** b **12.** -43 **13.** -13

LESSON 67

1. $4\frac{5}{11}$ **2.** $3\frac{1}{9}$ **3.** 1 **4.** $-\frac{3}{17}$ **5.** $15\frac{1}{4}$ **6.** -32 **7.** $-\frac{2}{3}$ **8.** 62
9. -25 **10.** -7 **11.** -5 **12.** -20 **13.** $-7\frac{1}{4}$ **14.** -27 **15.** 7
16. $3\frac{1}{9}$ **17.** $-\frac{10}{13}$ **18.** $-2\frac{2}{3}$ **19.** $-5\frac{3}{4}$ **20.** -82

LESSON 68

1. 10x + 6y **2.** 14x - 13y **3.** 13x + 2y **4.** 2x
5. 11x + 2 **6.** 6x **7.** d **8.** b **9.** d **10.** 4x - 6 **11.** 5
12. 11x + 10y **13.** 7x - 10y **14.** -2x - 2y

15. -5x + 3y **16.** 9x + 10 **17.** 5x + 8 **18.** c **19.** a
20. d **21.** c **22.** -x + 6 **23.** 3 **24.** 9x + 3
25. 11x - 5y **26.** 10x + 6y **27.** b **28.** 13x + 4y
29. no **30.** yes **31.** 11ab + 20a + 8
32. 9yz + 22y + 5 **33.** d **34.** d **35.** 9jk + 13j + 13
36. 9uv + 16u + 9 **37.** yes **38.** no
39. 11de + 20d + 10 **40.** 6wx + 10w + 13
41. 9mn + 10m + 7 **42.** 8yz + 20y + 6 **43.** c **44.** c
45. 3ab + 19a + 10 **46.** yes **47.** 10pq + 15p + 8
48. c **49.** yes **50.** no **51.** a **52.** a **53.** b
54. 10a + 4b - 12c **55.** -d + 19e + 8f
56. 27x + 3y - 26z **57.** $6e^2 + 5ef - 10ef^2$
58. $-9x^2 + 18xy + 4z^2$ **59.** $-3a - b - 2c - 4a^2$
60. $-7abc + 6a^2b^2c^2$

LESSON 69

1. -13 **2.** -25 **3.** b **4.** d **5.** a **6.** -43 **7.** 55 **8.** a **9.** d
10. -20 **11.** -20 **12.** d

LESSON 70

1. x = $1\frac{1}{7}$ **2.** x = -1 **3.** x = $-\frac{2}{3}$ **4.** a **5.** d **6.** x = 1
7. x = 1 **8.** x = $1\frac{1}{7}$ **9.** x = -3 **10.** b **11.** a **12.** b
13. x = 2 **14.** x = $\frac{25}{3}$ **15.** x = 64 **16.** x = 16
17. x = 20 **21.** N = -16 **22.** x = 10 **23.** x = -12

LESSON 71

1. -1 **2.** 2 **3.** a **4.** 3 **5.** b **6.** d **7.** -6 **8.** 3 **9.** 1 **10.** d
11. a **12.** a **13.** d **14.** 6 **15.** -2 **16.** c **17.** c **18.** -6
19. 4 **20.** -2 **21.** 2 **22.** 1 **23.** 5 **24.** -6 **25.** $\frac{2}{3}$ **26.** 10
27. d **28.** a **29.** d **30.** d **31.** $-\frac{7}{18}$ **32.** $-\frac{10}{17}$ **33.** -6
34. -5 **35.** -4

INDEX